Technological Change Collective Bargaining and Industrial Efficiency

Technological Change Collective Bargaining and Industrial Efficiency

PAUL WILLMAN

CLARENDON PRESS · OXFORD
1986

Oxford University Press, Walton Street, Oxford OX2 6DP
Oxford New York Toronto
Delhi Bombay Calcutta Madras Karachi
Petaling Jaya Singapore Hong Kong Tokyo
Nairobi Dar es Salaam Cape Town
Melbourne Auckland
and associated companies in
Beirut Berlin Ibadan Nicosia

Oxford is a trade mark of Oxford University Press

Published in the United States
by Oxford University Press, New York

British Library Cataloguing in Publication Data
Willman, Paul
Technological change, collective
bargaining, and industrial efficiency.
1. Technological innovations—Social
aspects 2. Labor and laboring classes
—Attitudes 3. Trade-unions
I. Title
338'.06 T173.8
ISBN 0–19–827262–6

Library of Congress Cataloging in Publication Data
Willman, Paul
Technological change, collective bargaining, and
industrial efficiency.
Bibliography: p.
Includes index.
1. Trade-unions—Great Britain. 2. Technological
innovations—Great Britain. 3. Efficiency, Industrial—
Great Britain. 4. Trade-unions—United States.
5. Technological innovations—United States.
6. Efficiency, Industrial—United States. I. Title.
HD6664.W56 1986 331.88'0941 86-5104
ISBN 0–19–827262–6

Set by Spire Print Services Ltd, Salisbury
Printed in Great Britain
at the University Printing House, Oxford
by David Stanford
Printer to the University

Collective bargaining has proved to be an excellent vehicle for the effective management of change

Technology and the American Economy, report of the National Commission on Technology, Automation, and Economic Progress, Vol. 1, 1966, 65.

Preface

My main objective in this volume is to try to explain the origins and the incidence of trade union resistance to technical change. Although there is some comparative material, and although on occasion there are some historical references, the book is primarily concerned with relatively recent events in the UK, since I was particularly concerned to examine the proposition that there is something distinctively unfortunate about the behaviour of British trade unions in the face of change which operates to the competitive disadvantage of unionized firms in the UK. This has always seemed to me to be an area in which more has been said than has been substantiated by the available empirical evidence.

A second objective arose from this. There is a sense in which students of trade union behaviour concern themselves primarily with dependent variables in that, notwithstanding the capacity of unions to affect the environment in which they operate, much union activity is in reaction to previous initiatives by employers or governments. Where this initiative is a proposal for or an instance of technical change, it becomes extremely important to understand something about the change itself and the intentions behind it if one is to deal with the fact that apparently similar technical changes provoke different responses on different occasions.

My second objective is thus to integrate the work of researchers in a number of different subject areas into a framework which attempts to understand changing labour relations within the firm; these subject areas include innovation studies, manufacturing policy, organizational behaviour, and business strategy, as well as the conventionally descriptive approach of academic industrial relations. The basis of this framework lies in the so-called 'new' institutional economics, and the book may be seen as an attempt to explore the usefulness of this sort of theorizing for those concerned with labour relations.

The book draws together the results of several research projects supported by different funding bodies. I am grateful for the support both of the Joint Committee of the (then) Science Research Council and Social Science Research Council and subsequently of the Economic and Social Research Council which enabled the collection of the data on which Chapters 7, 9, and 10 are based, and for the co-operation of the firms concerned. The comparative material on the USA was collected with the assistance of support from the Nuffield Foundation and a grant from the US Government. Drafts of the volume have been inflicted on members of the secretarial staff of Imperial College, Cranfield School of Management, and London Business School; I am particularly grateful to June Wardill and Pat Burge. I am grateful also to academic colleagues who have commented on all or part of the manuscript. These include participants at the various conferences and seminars at which parts of this volume have been presented over the last few years, and colleagues at Cranfield and London Business School. In addition, I am particularly grateful to Prof. D. E. C. Wedderburn and Prof. R. Martin. Finally, I owe thanks both to David Bell and Ed Sweeney for comments on the banking material and to David Buckle and several members of Austin Rover management for their help with the material on the car industry. Of course, the usual disclaimer applies: only I am responsible for any errors or misconceptions which remain.

PAUL WILLMAN

London Business School
July 1985

Contents

List of Figures

List of Tables

List of Abbreviations

ACAS	Advisory, Conciliation, and Arbitration Service
AESD	Association of Engineering and Shipbuilding Draughtsmen
AFL–CIO	American Federation of Labor and Congress of Industrial Organizations
AIF	annual improvement factor
APEX	Association of Professional , Executive, Clerical and Computer Staff
APS	Audited Plant Status
ASSET	Association of Supervisory Staffs, Executives, and Technicians
ASTMS	Association of Scientific, Technical, and Managerial Staffs
ATM	Autoteller machine
AUEW	Amalgamated Union of Engineering Workers
AUEW(E)	Amalgamated Union of Engineering Workers (Engineering Section)
BIFU	Banking, Insurance, and Finance Union
BJIR	British Journal of Industrial Relations
BL	British Leyland
BLMC	British Leyland Motor Corporation
BLS	Bureau of Labor Statistics
BMC	British Motor Corporation
BNA	Bureau of National Affairs
BPIF	British Print Industries Federation
BSI	British Standards Institution
BTDB	British Transport Docks Board
CAD	computer-aided design
CAM	Computer-aided manufacture
CBI	Confederation of British Industry
CNC	Computer numerical control
COLA	cost-of-living adjustment
CPRS	Central Policy Review Staff
CSEU	Confederation of Shipbuilding and Engineering Unions
DE	Department of Employment
DPE	Department for Professional Employees (AFL–CIO)
EEF	Engineering Employers' Federation
EETPU	Electrical, Electronic, Telecommunications, and Plumbing Union

EITB	Engineering Industry Training Board
EIU	Economist Intelligence Unit
FMCS	Federal Mediation and Conciliation Service
GAO	General Accounting Office
GM	General Motors
IAM	International Association of Machinists and Aerospace Workers
IDS	Income Data Services
ILWU	International Longshoremen's and Warehousemen's Union
IOB	Institute of Bankers
IRRA	Industrial Relations Research Association
IRRU	Industrial Relations Research Unit
ISIC	International Standard Industrial Classification
IT	information technology
ITU	International Typographical Union
LSP	London Scale of Prices
M&M	Mechanization and Modernization
MLH	Minimum List Heading
MSC	Manpower Services Commission
NALGO	National and Local Government Officers' Association
NASD	National Association of Stevedores and Dockers
NATSOPA	National Society of Operative Printers, Graphical and Media Personnel
NCTAEP	National Commission on Technology, Automation, and Economic Progress
NEDO	National Economic Development Office
NGA	National Graphical Association
NIESR	National Institute of Economic and Social Research
NLRB	National Labor Relations Board
NPC	National Ports Council
NUBE	National Union of Bank Employees
NUJ	National Union of Journalists
NUPE	National Union of Public Employees
NYSA	New York Shipping Association
OECD	Organization for Economic Co-operation and Development
PEP	Political and Economic Planning
PMA	Pacific Maritime Association
POEU	Post Office Engineering Union
PSI	Policy Studies Institute
R&D	research and development
SCPS	Society of Civil and Public Servants
SIC	Standard Industrial Classification

SLADE	Society of Lithographic Artists, Designers, Engravers, and Process Workers
SMMT	Society of Motor Manufacturers and Traders
SOGAT	Society of Graphical and Allied Trades
SPRU	Science Policy Research Unit
TASS	Technical and Supervisory Section (AUEW)
TGWU	Transport and General Workers' Union
TSB	Trustee Savings Bank
TUC	Trades Union Congress
UAW	United Automobile Workers' Union
USDAW	Union of Shop, Distributive, and Allied Workers
VDU	visual display unit

1

Technological Change, Trade Unions, and Efficiency

INTRODUCTION

THERE has recently been a resurgence of interest in the relationship between technological change in industry and the activities of trade unions. In part this has occurred because rapid developments in technology, particularly that based on the application of microprocessors, have led to a concern with rates of innovation as an influence on national economic performance. In part, also, the prospect of a relatively rapid phase of innovation has encouraged trade unions themselves to take part in a national debate about the uses and consequences of information technology, and students of trade union behaviour to speculate upon the possible impact of widespread change on the sphere of industrial relations itself.

In practice, the major concerns of trade unions hinge around the impact of technological change on employment levels, on the demand for skills, and consequently on trade union membership or bargaining leverage. However, another major concern is that trade union behaviour will exert a dampening effect upon both rates of and returns to innovation. The evidence of highly visible resistance to change in industries such as docks and national newspapers supports the views of several commentators (e.g. Pollard 1982; Olsen 1982) that trade union behaviour may frustrate or discourage innovation within the UK, and is consistent with a body of work which suggests that poor labour relations contribute to low productivity growth rates in this country (Caves 1968, 1980; Pratten 1976; Prais 1981). Pollard, for example, suggests that many unions will accept technical change 'only after bitter struggles', and that innovations are held up for many years by trade union resistance (1982, 107).

This argument, that trade unions in the UK are obstacles to successful innovation, merits the closest attention. The first point to make is that, if unions do act in this way, their actions stand in stark contrast to their policy statements: throughout the post-war period at least, and particularly in their most recent statements on the subject, the TUC and member unions have consistently promoted the view that rapid innovation in industry is desirable and, furthermore, that it should be encouraged by government (TUC 1965, 1979). The second point is that, on the face of it, there is little evidence of trade union resistance: many commentaries on resistance lack an empirical basis. Moreover, an analysis of principal stoppages due to industrial disputes in the UK in the last twenty-five years reveals that very few have occurred explicitly over the introduction of technological change: the most overt form of resistance is rare.

The third point concerns the theoretical basis for the argument. Within conventional economics, it is axiomatic that trade unions as institutions are sources of allocative inefficiency. It is but a short step to the argument that the maintenance of inefficiency and resistance to or obstruction of change are linked: in fact, this link is established in the standard analysis of the phenomenon of 'feather-bedding' (see, for example, Addison and Siebert 1979, 315–16). However, institutional economists such as Williamson (1975) argue that under certain circumstances trade union activity may actively contribute to internal organizational efficiency. The literature on idiosyncratic exchange is also at the root of the argument by members of the so-called 'Harvard' school (e.g. Freeman 1976) that there may be a positive relationship between trade-unionism and productivity based on the union's ability to assist the organization to reverse 'decline' in responding to changing circumstances.

Since the contention that trade unions have a positive impact on productivity and efficiency is essentially at odds in the longer term with the argument that they prevent or obstruct change, the work of such institutional economists supports the instincts of academic researchers into industrial relations to resist simple generalizations about trade union behaviour and its consequences. Even if there is a simple negative relationship between trade unions and efficiency, the behaviour of some unions may encourage greater inefficiencies than that of others. Similarly, the disposition and capacity of organized labour in the UK to resist technological change may be unevenly distributed.

THE IDENTIFICATION OF TECHNOLOGICAL CHANGE

Some of this confusion may be dispelled by looking more closely at what one means by technological change. It is customary, for example, to distinguish between innovations in manufacturing processes and product innovation. In practice, the division is not clear cut, since product innovation may occasion new processes and vice versa: Abernathy (1978, 82), for example, outlines the relationship between changing processes and the product life cycle for the US car industry, illustrating the dependence of product development upon both changes in market structure and the availability of new processes.

However, a more central, if slightly naïve, division for our purposes is simply that between changes which are important for industrial relations and those which are not. Consider the following ascending sequence of changes, all of which are considered, for the moment, at firm level: their possible consequences for industrial relations are illustrated.

(i) Small-scale changes in materials or product with constant production technology: for example, the sorts of changes in the engineering industry described by Brown (1973) which might occasion, in the local negotiation of a new price or time, changes to working practices or earnings under payment by results.

(ii) Technological changes which do not directly affect the production process but which might be the subject of bargaining: for example, changes to performance monitoring systems such as those on shop-floor data capture which featured in negotiations between the UAW and Ford in the late 1970s in the USA.

(iii) Major changes to production processes which lead to a reduced demand for labour, but which require approximately the same sorts of skills: for example, the introduction of automated body assembly as at British Leyland (Willman and Winch 1985).

(iv) Major changes to production processes which alter both the scale and the nature of demand for labour, and remove old skills altogether: for example, the use of photocomposition in the national newspaper industry (Martin 1981).

This list does not constitute a classification, nor does it exhaust the possible types of change, but it does serve to illustrate several points. First, the examples illustrate that many of these types of change need not trigger events which industrial relations specialists would consider to be of interest. Many industries have recurrent small-scale change and performance monitoring systems which are not the subject of bargaining: major process innovations such as the computerization of banking operations have occurred without, to date, conflict or negotiation, although they may subsequently have consequences for collective bargaining. Secondly, changes which *do not* involve new technology in any direct way may occasion either greater productivity benefits or greater trade union resistance than those which do. For example, the success of productivity bargaining at Fawley involved no new technology (Flanders 1964), nor did the disputes about flexible rostering on the railways, or the reform of collective bargaining at BL Cars.

This line of reasoning supports the proposition that there is no simple relationship between the size of any technological change (measured, for example, by the cost of new capital equipment) and resistance to it. Moreover, changes which go through without negotiations in certain sectors of the economy would be bargained in others. As Slichter *et al*. noted of the USA, one must look at the nature of the union *and* the condition of the enterprise in order to explain trade union responses (1960, 345).

DECISIONS TO INNOVATE

However, the analysis of trade union responses focuses upon what is primarily a dependent variable. Although expectations about trade union responses may condition the innovative behaviour of companies or trade union influence on decision making may exert pressure on the form or speed of change, essentially the primary issue to consider is the purpose for which investment in new technology is considered.

Here it is possible to spell out at least seven different motiviations which, singly or in combination, might underlie proposals for change. First, companies might consider change in order to reduce labour costs, either by directly reducing the numbers of people employed or by reducing the skill content of jobs and hence the supply price of future job-incumbents. By contrast, certain companies may experience skill shortages, and automate processes in order to avoid difficulties in the recruitment or retention of scarce skills.

Another set of motives for innovation concerns the market for products rather than the market for labour. Companies may innovate, for example, in the adoption of computer-aided design in order to reduce lead times on design or in production. They may adopt process innovations, for example the automation of paint spraying in the car industry, to permit an improvement in some aspect of product quality. Thirdly, they may innovate to expand output of a given product or to produce additional products or services.

The final set of reasons is less homogeneous. Companies may adopt new techniques which are, strictly speaking, administrative rather than either process or product changes: for example, the improvement of management information systems, including performance assessment and the reduction of inventory by the adoption of computerized stock control. Finally, companies may buy new capital equipment simply to replace worn-out, obsolete, or fragile equipment.

This range of possible motives ignores for the moment the potentially large class of cases where companies simply engage in imitative behaviour in the course of the diffusion of a process innovation. However, the range presented here is again sufficiently broad to discourage any view of innovative behaviour as a homogenous set of events. Although much depends in addition on the details of any particular innovation, the response of unions to change which is a direct assault on employment levels is likely to be very different from their response where additions to the product range are envisaged.

THE IDEA OF RESISTANCE TO CHANGE

The final issue to be considered at this stage is the idea of trade union resistance to change. Again, this general idea can be broken down into a number of propositions, only some of which are testable. The most important of these are as follows:

(i) That companies considering new investment in the UK are dissuaded from doing so by considerations which have to do primarily with the behaviour of employees organized in trade unions.
(ii) That companies already operating within the UK do not consider innovation except *in extremis* because of their expectation that labour resistance will cause unacceptable costs to be associated with innovation.
(iii) That companies fear a phase of resistance, possibly including strike activity, around the time the innovation is implemented and include the costs of this phase in their calculations about the decision to innovate.
(iv) That companies approach innovation expecting to pay wage

rises to those operating the new equipment before the gains of innovation can be guaranteed.

(v) That companies expect the benefits of innovation in the UK to be reduced in the long term by the maintenance of restrictive practices and excessive manning levels by trade unions.

These propositions need not exhaust the range available to illustrate that there are a number of difficulties involved in evaluating the argument. For one thing, in assessing union resistance, instances where technological changes do *not* occur become as important as those where they occasion strikes or visibly remove restrictive practices. The UK might, for example, suffer from low productivity because trade unions discourage new investment. Moreover, some forms of resistance, for example the bargaining of payment for change, might in the longer term actually accelerate investment which substitutes new equipment for expensive labour.

THE PROBLEM

In summary, then, the relationship between trade unions and technological change is difficult to describe in general terms. The straightforwardly obstructive role described by conventional economics is difficult to accept not only because of developments within economics itself but also because closer inspection of the posited relationship reveals a relatively complex set of questions only some of which are answerable.

Moreover, answers to this latter sub-set require that one consider the motives behind and extent of change as well as the structure and practice of trade-unionism within particular firms and industries. The central questions are:

1. Why do some instances of technological change occasion resistance from trade unions whereas others do not?
2. What determines the forms such resistance takes?

In posing these questions I am clearly focusing attention away from non-union environments as well as from situations such as those instanced above, where the prospect of change is not mooted because of the fear of an aggressive union response. This volume is intended to fill part of the gap identified by Bell, namely the lack of a comprehensive study 'indicating where, when, how and why the organisation, power, attitudes and behaviour of labour have excited a peculiarly British set of negative influences on the process of technological change' (1983, 3). The structure and argument are as follows.

Chapter 2 looks at the development of trade union policy towards technical change by analysing the course of the two post-war 'automation' debates in the UK and USA: these debates reveal a complex relationship between rates of innovation and trade union policy which implies the need to look more closely at the evidence for trade union obstruction or discouragement of change. The evidence is reviewed in Chapter 3: it is equivocal, but appears to point to a particular problem in certain industries, and particularly in large unionized companies. Chapter 4 then seeks to understand the process of innovation in such companies: using the approach developed in a number of publications by Abernathy and Utterback, it sets the sequence of product and process innovation in the context of corporate competitive strategy; given the development of such strategies over long periods of time, the approach suggests that institutions for the management of labour must also evolve.

An approach to the analysis of this evolution is presented in Chapter 5. Relying on the literature on idiosyncratic exchange, it suggests changes in the form of labour contract specifically related to changes in the production process and the product-market. In Chapters 6 and 7 this approach is used to structure analysis of the three industries which in post-war Britain have accounted for most of the working days lost in disputes over technical change—docks, national newspapers, and motor vehicles. In the first two, considerable parallels occur in technical and institutional change which moreover further parallel developments in those industries in the USA. In motor vehicles however, such international comparisons break down in the absence of similar labour institutions.

Subsequent chapters focus on more recent events in industries affected by micro-electronics. Chapter 8 argues that the most difficult issues concern the acceptance of change in mature industries which have to accommodate cost-cutting process innovations based on micro-electronics. In Chapters 9 and 10 two such industries are looked at in greater detail through detailed case studies of two large firms. Chapter 11 concludes with a general discussion of the relationship between innovation and trade union activity.

2

Two Automation Debates

INTRODUCTION

I HAVE already noted that, although trade union policy statements have tended to emphasize the desirability and necessity of technological change, at least one school of thought suggests that in practice the trade unions impede such change. The most straightforward resolution to this paradox is simply to say that in the UK the decentralized structure of trade union organization prevents the implementation of policy initiatives emerging from the centre: whatever is planned at national level, the sectional interests of those who *can* resist change or appropriate its benefits will override. However, this response raises fundamental issues to do with the nature of trade union activity in the UK which in turn have to do with the question of why the UK might appear disproportionately to suffer from sectional behaviour.

This chapter will focus on the development of trade union policies towards new technology, both in the UK and in the USA. I shall try to show both that there exists substantial continuity in the two countries in such policy development, and that the AFL–CIO and TUC have equally come to depend upon appeals for government action. However, whereas in the UK one can identify two 'automation debates' within the trade union movement, the USA has, thus far, not generated a second: I shall be concerned to analyse the factors underlying this. The chapter will thus initially describe developments in the UK and USA, and subsequently turn to an explanation of differences.

THE DEVELOPMENT OF TUC POLICY

In the UK, the concern with technological change is evident in discussions at Congress as early as 1947. However, the content of these discussions was, by present-day standards, quite narrow. As Benson and Lloyd note, 'the issue of new production technology was dealt with simply as one of job displacement' (1983; 71). Trade union concern was also illustrated in the re-establishment of the Science Advisory Committee—a body intended to promote technological change in industry—involving both TUC nominees and members of the British Association for the Advancement of Science. However, at this stage, a number of concerns which were later to become important, such as union involvement in the design of change, the vetoing of change, and the involvement of government, were absent. Greater emphasis was given to the need to assist technological progress.

This emphasis continued through what Benson and Lloyd have termed the 'first' automation debate (1983; 73) in the mid-1950s. In 1955, the TUC General Council produced a 'Memorandum on Automation' which was adopted as policy by Congress the following year and was to provide, through successive revisions in 1965 and 1970, the basis of the TUC's policy on the issue until 1979. The mood and central thrust of the document are captured in the following:

> In the post-war period there have been no serious technological redundancy problems which have been beyond solution by the firms concerned or locally. Difficulties might arise, however, if the rate of automation accelerated suddenly, depending again on whether the application was in new and developing industries or substituting established production methods in existing industries.
>
> The major job of the trade unions will be to keep automation within the field of industrial relations; countering assumptions that working conditions, earnings opportunities, pace of working and other matters affecting workpeople can or should be arbitrarily fixed by machines or technicians—that there is no opportunity for joint negotiations and the expression of trade union points of view on questions connected with industrial efficiency and development. (TUC *Report* 1955, 249.)

At this stage, there was very little reliance on proposals for government action. In fact, in 1955 a motion from NUPE calling for public ownership as a control over the rate and direction of technological change was defeated, while the following year a motion from ASSET proposing national planning through a tripartite Board for Automation which would ascertain when automation was 'in the interests of the nation', was similarly rejected (TUC *Report* 1956, 370). The emphasis remained primarily on the establishment of local devices to deal with local problems. The General Council's *Report* states:

> From their analysis . . . the General Council are not convinced any more than last year that there is any justification in concluding that the spread and

extent of automation . . . is going to be as alarming as is sometimes forecast, notwithstanding the possibility of highly localised significant developments and changes. (TUC *Report* 1956, 518.)

By 1958, TUC policy had thus formed around a concern to encourage technical change in public statements and through limited involvement with various advisory groups at national level, relying on local negotiations to provide job-security safeguards. The policy was to develop in a number of respects over the next few years. A concern with office automation first arose in 1956, and became more prominent in later years.[1] Reports of special conferences on technology frequently featured as part of the General Council's annual report. However, the policy did not fundamentally change, and the biennial survey of affiliated unions by the General Council to monitor the impact of automated techniques gave little cause for disquiet. In 1960, the General Council argued that

it would be unwise to over-dramatise automation or to attempt to isolate it from other significant developments. . . . In conditions of full employment the problems of automation have been shown to be more manageable than those relating to the rapid contraction of industries such as cotton and coal. (TUC *Report* 1960, 154.)

Moreover, no grounds for revising the 1955 policy impressed themselves on the General Council in the early 1960s. It was still felt that too little change, rather than too much, was the greater fear (TUC *Report* 1963, 251). However, this was not a matter for consensus, and several affiliates sought a more radical stance. For example, in 1957 the AESD tabled a motion seeking national planning on technological change with effective trade union participation, while in 1963 the sheet-metal workers tried to refer back those paragraphs of the General Council's *Report* which argued for the beneficial consequences of change: both attempts failed.[2]

Although the TUC position may now appear relaxed, it also appears justified by conditions at the time. Slichter *et al.* argue that trade union policies towards technological change depend on four main factors. The first is the proportion of union members affected by the change, the second is the economic condition of the industry or enterprise, the third the nature of the technological change, and the fourth its stage of development. Most frequently trades unions will respond to change with 'willing acceptance': they move to more resistant stances, such as seeking 'adjustment' mechanisms or offering outright opposition, only where the pervasiveness of the change or the economic circumstances surrounding it produce substantial threats to job security (1960, 342–71). I shall deal with this in more detail later. The central point here is that both the buoyant economic climate of the late 1950s and the absence of any pervasive set of innovations—such as, for example, those related to microprocessor technology which might affect the bulk of the TUC's membership—reduced the anxiety voiced at Congress about the scale of the displacement problem.

Both of these factors were to change by the late 1970s, but the impetus for the first shift in policy emphasis came from another source. In 1964, the General Council decided on the basis of its survey of affiliates that a revision of its automation policy was called for. Although, as Benson and Lloyd note, this revision concerned itself with essentially the same policies, noting only the potential for accelerated change with a wider impact, its adoption by Congress as *Automation and Technological Change* (TUC 1965) was followed by discussions with ministers of the newly elected Labour Government 'to secure information on the Government's policies and intentions in regard to science and technology, including the part that trade unions might be expected to play' (TUC *Report* 1965, 136). The concern was still to promote innovation, and the widening technological gap between the UK and USA was a particular cause for concern (TUC *Report* 1969, 467). By the time of the second, 1970, revision of the approach, however, the policy of willing acceptance of technological change by Congress is tempered by a concern to ensure that employers and government both play their part:

> the prospect of more rapid and more extensive technological advance . . . is welcomed by the TUC provided that the responsibilities which accompany it are accepted and discharged by the Government and employers as well as by the trade union movement . . . the Government's responsibility is to co-ordinate progress and so to plan the nation's affairs as to secure the optimum rate of advance and the maintenance of full employment. (TUC *Report* 1970, 38.)

The General Council statement on science in 1971 was overtly critical of the failures of government policy on innovation (TUC *Report* 1971, 232–6), while in 1972 a motion was passed arguing that the consequences of change 'are only acceptable in the context of planned economic growth, and as a means to greater leisure and social activity, not as a means of reducing the workforce and introducing other anti-social effects' (TUC *Report* 1972, 198). However, rather suddenly the issue disappears from Congress agendas: between 1973 and 1978 there are no major motions or serious discussions. Yet, when the concern about automation and technical change re-emerged in the late 1970s, the role of government in the promotion of change and the planning of the economy was a central issue from the outset.

EMPLOYMENT AND TECHNOLOGY

The second automation debate within the trade union movement began in 1978. At Congress that year the POEU moved a resolution calling for wide-ranging government action both to encourage technological change and to ensure that any adverse social or economic consequences were minimized. This was to be achieved by providing for an expansion of job opportunities and for extensive retraining to prevent mismatches between available jobs and existing skills in the labour-market.[3] In addition, the resolution initiated a study into the whole area of technological change to be conducted by a specially constituted Employment and Technology Committee. This committee, composed of representatives from major interested unions and supported by TUC staff, produced an *Interim Report* for a special conference on technology in May 1979, and a *Final Report* accepted by Congress later that year.[4]

The reports, and the activities of the Employment and Technology Committee, provided the basis for the most extensive discussion of technological change within the union movement in the post-war period. Many unions produced policy documents which to a greater or lesser extent mirrored the TUC *Report*, and an extensive educational campaign was mounted.[5] However, the debate was relatively short-lived. The election of May 1979 tended to shift the priorities accorded to various policy areas. By the 1980 Congress, Conservative legislation on trade union behaviour and the incoming government's economic policy were important: although two motions mentioned new technology in the context off concern with economic policy and the length of the working week, the focus on the 'microprocessor revolution' was decidely muted.[6] By the 1981 Congress, the issue had moved even further down the agenda. Congress focused on unemployment, economic policy, and the forthcoming battle for the deputy leadership of the Labour Party.

The policy on technology had received a substantial setback in the intervening year. The TUC had attempted, unsuccessfully, to secure CBI agreement to a joint statement on the regulation of technological change which would have involved joint consultation, sharing of benefits, the provision of job-security guarantees, and a commitment to retraining. Despite agreement with the Confederation itself, the CBI membership rejected interference with established provisions at firm and industry level (CBI 1980; Bamber and Willman 1983). By the time of the 1981 Congress, technological change featured in the General Council's report,[7] rather than in general debate. Since both the CBI and the government seemed disinclined to become involved in tripartite action, further initiatives seemed irrelevant.

The central feature of both the 1978 conference and the subsequent *Interim Report* is that by the 'second' automation debate the pendulum had swung completely away from the relaxed attitude of the 1950s involving reliance on local adjustment to change. The resolution from the POEU makes no mention of collective bargaining: the *Interim Report* outlines a programme for trade union action involving concerns for industrial democracy and for the signing of 'new technology' agreements, as well as an educational programme to develop awareness of the threats of technology. However, the emphasis is still that 'the problems associated with economic and technological change will not be resolved without an industrial strategy operating at all levels of the economy' (*Interim Report* 1979, 49).

In outline, the argument runs as follows. Historically, high productivity has been associated with high output growth, good trading performance, and steady or rising employment. Rapid technological change could therefore set the economy on the path to higher output, lower unit costs, and full or near-full employment. However, three basic conditions need to be fulfilled before innovation can be a success in these terms. First, the government must demonstrate a commitment to an industrial strategy involving directed investment, the maintenance of high levels of demand, and the monitoring and containment of import penetration. Second, the public services must be expanded to absorb the employment surplus. Third, there needs to be international co-operative action by governments to promote expansion of world trade. In addition to all of this, a detailed manpower policy is spelt out, involving co-ordinated action by universities, polytechnics, industrial training boards, and the Manpower Services Commission, to provide a 'comprehensive' employment service.[8]

Substantial changes are also proposed to redundancy payments schemes, involving continuous rather than lump-sum benefits payments.

The second automation debate was thus initiated with a reliance on government action which involved treating the problem of technological change as an area to which the Keynesian analysis characteristic of successive *Economic Reviews* could be applied. Essentially, the basic protection for trade union members threatened by technological change was to be provided by policies which emerged from dialogue between the TUC and a sympathetic government. However, such a view could not survive unscathed after the disappearance of such a government, and by the 1979 Congress delegates were being asked to accept a *Final Report* which involved a number of changes.

Some of these were simply elaborations of positions briefly stated in the *Interim Report*. For example, in the later version there is increased emphasis on the problems faced by women in employment and on their training and educational requirements: this reflected recognition of the severe impact which developments in microprocessor-based office technology might have on the restricted range of clerical occupations in which much female employment is concentrated. However, the most substantial changes reflected a growing awareness that the requisite government action to encourage painless change was unlikely to be forthcoming under a Conservative administration. Whereas the *Interim Report* discussed details of the putative government strategy, the *Final Report* came strongly to the defence of the public sector as an essential component in economic growth. Suggestions that technological change was necessarily to be associated with a given level of unemployment were rejected as 'deterministic' (*Final Report* 1979, 13, 29).

However, the most basic change was the addition of a check-list of items to be negotiated under circumstances of technological change. This check-list is by now familiar, having been reproduced in a relatively large number of academic and trade union publications (see Table 2.1). The point to be emphasized here is that, whereas one might suggest that the proposals for government action to facilitate and lubricate the process of change represented a qualitative advance on the policy stances of 1955 and 1965, the collective bargaining strategy remained essentially the same. As in the first automation debate, prior consultation, change by agreement, retraining provisions, guarantees of job security, and reductions in working hours all featured in the TUC check-list for the second. The principal addition was the suggestion that negotiations should seek to influence the selection or design of new equipment.

In summarizing the development of TUC policy since the war, it is probably best to separate out those proposals which focus on collective bargaining and those which seek to promote wider action. Put simply, the former have been qualitatively very similar in the two

Table 2.1. TUC Check-list on New Technology Agreements

1. Change must be by agreement: consultation with trade unions should begin prior to the decision to purchase, and status quo provisions should operate until agreement is reached.
2. Machinery must be developed to cope with technical change which emphasizes the central importance of collective bargaining.
3. Information relevant to decision making should be made available to union representatives or nominees prior to any decision being taken.
4. There must be agreement both on employment and output levels within the company. Guarantees of job security, redeployment, and relocation agreements must be achieved. In addition, enterprises should be committed to an expansion of output after technical change.
5. Company retraining commitments must be stepped up, with priority for those affected by new technology. Earnings levels must be secured.
6. The working week should be reduced to 35 hours, systematic overtime should be eliminated, and shift patterns altered.
7. The benefits of new technology must be distributed. Innovation must occasion improvements in terms and conditions of service.
8. Negotiators should seek influence over the design of equipment, and in particular should seek to control work or performance measurement through the new technology.
9. Stringent health and safety standards must be observed.
10. Procedures for reviewing progress, and study teams on the new technology, should be established.

Source: TUC 1979.

debates: the principal exception is the attempt to influence equipment design. However, the period since 1955 has been characterized by a growing awareness of the need to promote governmental action in this policy area, in part because of the inadequacy of collective bargaining mechanisms alone. This shift of emphasis can be related to changes in the four major influences on union policy noted earlier: the scale of the change, its stage of development, the state of the economy, and the nature of the trade union movement.

CONTINUITY AND CHANGE

The second automation debate was triggered by a growing awareness of the implications of micro-electronic technology which included a programme of government support for its adoption (Benson and Lloyd 1983, 167). However, it began at the end of a period when the increase in basic process innovations was relatively rapid from quite a low base. Figure 2.1 reproduces Freeman *et al.*'s presentation of basic innovations in the UK since 1920 (1982, 52): it is apparent that the initiative of 1955, which had provided the basis for TUC policy for the two decades prior to the start of the second debate, occurred during a similar upswing in the appearance of radical process innovations. TUC policies, then, may have reflected innovative developments within manufacturing industry quite accurately: the rate of diffusion of such change is, of course, a different matter.[9]

However, although there may be similarities in the rate of increase of process innovations between the two periods, economic circumstances clearly changed. This may well have been related to technical development. Freeman *et al.* remark of the post-war UK that

> In the early period of a long boom, the emphasis is on rapid expansion of new capacity in order to get a good market share and this investment has a strong positive effect on the generation of new employment. As the new industries and technologies mature, economies of scale are exploited and the pressures shift to cost-saving innovations in process technologies. Capital intensity increases and employment growth slows down or even stops altogether. (1982, 21.)

This logic does seem to apply to UK manufacturing. Clark (1979) shows that manufacturing investment is associated with employment contraction rather than expansion from 1964 onwards. Similarly, Rothwell and Zegveld suggest a shift from expansionary to rationalizing investment over the same period. They conclude that

> fundamental structural changes have taken place in the relationship between manufacturing output and manufacturing employment, and between investment in manufacturing and manufacturing employment, during the thirty-year period 1950 to 1980. (1985, 32).

The stage of development of a technology is thus of considerable importance, and change in mature industries is likely to be more threatening to employment levels than new product development.

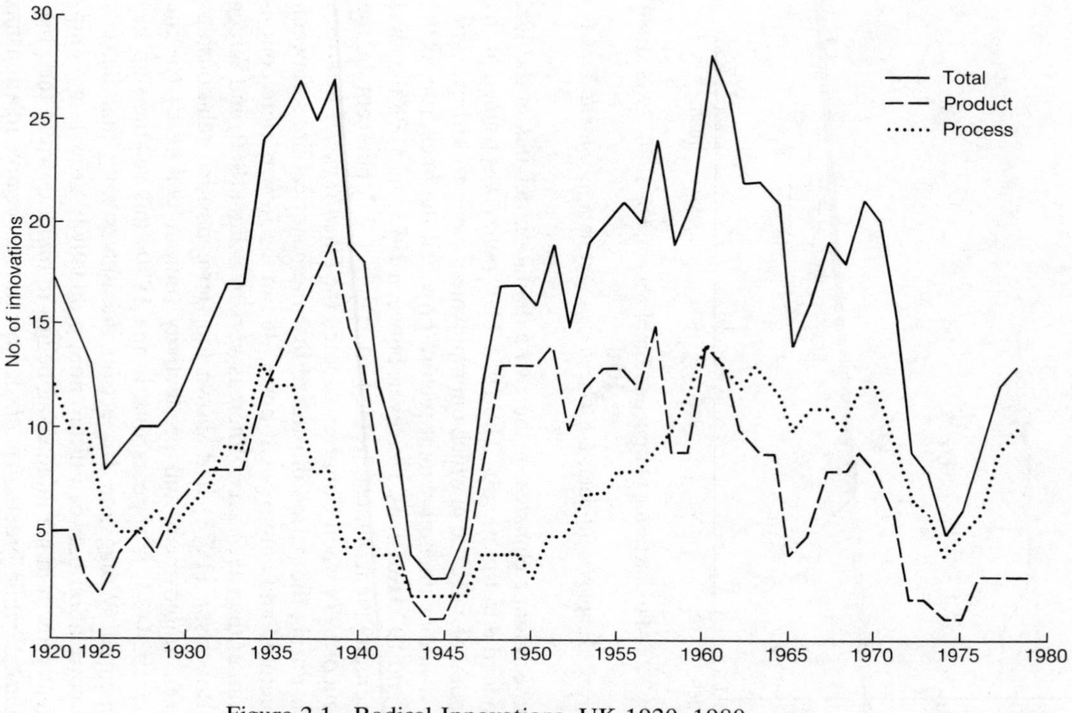

Figure 2.1. Radical Innovations, UK 1920–1980
Source: Freeman *et al.* 1982

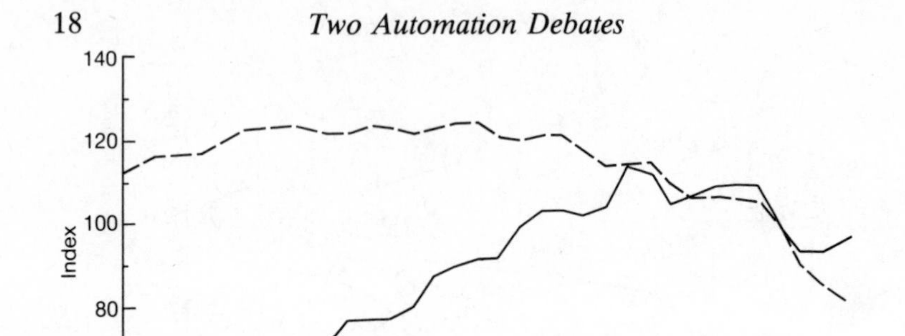

Figure 2.2. Manufacturing Output and Employment, UK 1950–1983 (1980 = 100)

Sources: Employment data, *DE Gazette*; output data, *Economic Trends*

The economic climates of the two automation debates in the UK did indeed differ markedly. The 1955 TUC policy was framed at the mid-point of a decade in which unemployment never rose above 3 per cent; a similar unemployment picture provided the backdrop to the 1965 and 1970 revisions. However, between 1973 and 1978 the level of unemployment almost trebled to stand at 5.7 per cent in the autumn of 1979 when Congress adopted the *Final Report*. Moreover, events during the period illustrated that the historical links between productivity and employment postulated in the *Report* were contingent rather than necessary. Whereas work by Salter (1960) and Wragg and Robertson (1978) had shown long-term positive relationships between employment and productivity (output per head) for the period 1924–73, it appears that in the 1970s this relationship has disappeared. Wragg and Robertson themselves note that Salter's close correlation between differential productivity growth and employment growth in manufacturing begins to break down in the post-war period, while Freeman *et al*. demonstrate a *negative* relationship between employment and productivity in this sector, and a weakening in the relationship between changes in employment and manufacturing output between 1973 and 1979 (1982, 134, 151, 153). As Figure 2.2 shows, the overall economic climate of the first automation debate (1955–65) was much more favourable than that which preceded the second.

For the framers of TUC policy in 1979, these trends reinforced the concern for economic reflation. In particular, the rapid increase in output per person employed, in the number of process innovations, and in the level of unemployment after 1975 presented a set of economic relationships rather different from those of the previous thirty years. Since, historically, policy had never been concerned with outright opposition to the productivity benefits attributed to technical change but rather with the possible displacement effects, the solution to the problem of technological unemployment was seen in terms of those arguments against demand deficiency which enlightened TUC economic policy more generally. Once more, this required government action.

The fourth and final factor underlying this change of policy emphasis was the changing nature of the TUC itself. Two rather different sorts of change are relevant. The first concerns the changing nature of the TUC's relationship to government, which is of interest given the increasing role proposed for government in policy statements on technological change. The second concerns the changing occupational composition of Congress which, I shall argue, has influenced the qualitative change noted in the collective bargaining strategy itself. The essence of the first argument is that the TUC in 1978 was much more favourably disposed towards reliance on the good offices of a Labour government which was party to the Social Contract than was its predecessor, operating under a Conservative administration in 1955. Moreover, its *internal* authority over affiliates was such that it was much more likely to outline a comprehensive bargaining strategy for them. On the first point, Martin notes that consultative relationships with Conservative governments in the 1950s were by no means poor. Indeed 'there was little to distinguish the Conservative governments of 1951–64 from the Attlee government, so far as their consultative relationship with the TUC was concerned' (1980, 301).

Yet the level of influence exerted by the TUC in the early 1970s, particularly through the TUC–Labour Party Liaison Committee, represented a substantial increase in external influence. Whereas under the post-war Labour government the TUC had tended to defer to the needs of the government rather than vice versa, and under the 1964–70 Wilson administration the relationship had tended to be fragile, the 1974–9 relationship was rather different (Martin 1980, 294–310). As Clark *et al.* note:

> At a time when the new Labour government was not insisting on significant concessions or immediate obligations from the TUC, many union leaders believed that they had established important new channels of influence on policy. They were involved in tripartite bodies with extensive executive functions, they were engaged in regular contact with cabinet ministers on the TUC–Labour Party Liaison Committee, and several of the most urgent demands of trade union members had been accepted as government policy. (1980, 64.)

It is of interest to note that this influence extended as far as attempts to *encourage* rather than *discourage* investment. The TUC had proposed to the Wilson Committee investigating the financing of industry and trade that an Industrial Investment Agency using North Sea oil revenues be established to improve the performance of manufacturing industry (TUC *Report* 1977, 85). Moreover, thirty-six sector working parties of the tripartite NEDC considered investment strategy as part of their brief.

Along with this growing external influence went a marked increase in internal authority. Beginning perhaps with the involvement in mediating unconstitutional stoppages in 1969, and gaining ground with the co-ordination of bargaining via the industry committees in the early 1970s,[10] the TUC exercised the most substantial control during the early part of the Social Contract. In 1974–5, as Clark *et al.* note, 'the TUC developed an operational function in collective bargaining that was unprecedented, and in direct opposition to strongly held beliefs in the autonomy of affiliated unions in wage bargaining' (1980, 64). The contrast here with the lack of power and influence over the pay-bargaining behaviour of affiliates in the 1950s has been noted by several observers: in Thomson's view, it renders Windmuller's conclusion that the TUC was among the weakest of national trade union confederations obsolete (Winchester 1979; Thomson 1979; Windmuller 1975).

The other major change was in the occupational composition of the trade union movement. Between 1955 and 1979 the number of white-collar trade-unionists increased substantially. In 1948, 22.6 per cent of union members in the UK were in white-collar employment; by 1978 this had grown to 40 per cent. The change has been extremely important for the development of trade union policy on technological change. Beginning with a concern within the Non-Manual Workers Committee about the impact of computerization in the 1960s, white-collar unions, particularly ASTMS, APEX, and

NALGO, moved to the forefront of policy development and implementation in the second automation debate (Robins and Webster 1982; Willman 1983b).

There are perhaps three broad reasons for this. First, the basis of this wave of technological change in microprocessor-based information technology offered the prospect of a breakthrough in the areas of information processing, storage, and retrieval which constitute a substantial part of clerical activity. Secondly, given the cheapness of information-processing power and the relative undercapitalization of office work, rates of penetration of office-automation devices over the medium term might prove to be too rapid to be absorbed through natural wastage or non-recruitment. Thirdly, white-collar unionization in the UK has extended to cover occupations which have a direct concern with the development of new technology. This has been true of ASTMS and APEX whose membership has extended to software engineers, computer programmers, and systems analysts in both manufacturing and service sectors.

This change in the nature of trade union membership may account for a distinctive addition to TUC policy in the second automation debate. It will be recalled that *Employment and Technology* argues that

> It is at the *design stage* for technological system that decisions will be taken that effect the technology's influence and control over those who work with it. Negotiators should therefore seek full involvement at this stage. (TUC 1979, 69.)

The suggestion is echoed in a number of publications, particularly those of ASTMS (1980). It presupposes the availability of expertise within the union both to question proposals for change and to offer alternatives which may be a characteristic of certain white-collar, rather than manual, organizations.

In summary, then, TUC policy, far from being one of opposition to change, has consistently been concerned to encourage innovation; however, this commitment has not been unqualified, but has rested on the development of collective bargaining strategies for adjustment and, latterly, on a concern to press governments to establish the 'correct' economic environment. The particulars of policy can be related to technological and economic variables, as well as to the nature of Congress membership.

This view of the costs and benefits of change was not necessarily universally held. Although most unions took a similar view to that of the TUC, and a review of policy statements concluded that 'not one trade union could be found which argued that technological change could be opposed outright . . . most favoured its development' (Robins and Webster 1982, 13), there still remained the issue of shop-floor acceptance. Written before the May election, the 1979 *Economic Review* pointed out that 'the TUC General Council are fully committed to the promotion of technological innovation and change. But the question is asked whether trade unionists at plant level are equally committed' (1979, 23). The TUC commitment was, of course, contingent upon certain forms of government action and collectively bargained safeguards for those affected. Without such action and safeguards, General Council commitment to change was doubtful. So much more doubtful, therefore, was the commitment of union members at plant level: I shall return to this below.

THE AUTOMATION DEBATE IN THE USA

A discussion of the automation debate of the early 1960s in the USA provides a useful basis for assessing the argument so far: if this argument has validity, it ought to be possible to show both that a resort to government assistance was a feature of AFL–CIO policy and that the failure of a 'second' debate in the USA is related to differences in the four main factors I have discussed so far.

A concern with automation, particularly in the public sector, emerged at the inaugural Convention of the AFL–CIO in 1955.[11] However, the scale of debate at the subsequent 1957 Convention was more substantial, with a resolution being carried calling for continuing study of the social and economic impact of new technology, for effective collective bargaining over change, and, at the outset, for legislative programmes to protect job security.[12] In 1959, and again in 1961, the Resolutions Committee proposed motions which recognized the particularly rapid changes occurring in mining, steel, and railroads. The 1959 motion specifically welcomed technological change, provided that it was to be used to improve working and living conditions and to be implemented after collective bargaining.[13] At both conventions the emphasis lay on collective bargaining *and* government action, the 1961 resolution urging

> federal, state and local governments to recognise, cooperate in and carry out their responsibilities towards meeting the national challenge that far exceeds the abilities or responsibilities of labor and management to solve.[14]

However, the most protracted debate occurred at the 1963 Convention, with the passing of a resolution calling for wide-ranging action, most of which relied in turn on action by federal or state legislatures. Resolution 222, recognizing the potential benefits of technological change and the need for the development of collective bargaining provisions, called also for: a Presidential Commission on Automation to look at the gamut of problems associated with job displacement; a technolgocial clearing-house to provide an 'early warning system' for the advent of change; measures to promote economic growth; improvements to unemployment compensation; an employment service to match individuals to jobs; and a federal information and guidance service to assist those displaced.[15] A further, and perhaps more radical, motion in 1965 called in addition for job opportunities for all those willing and able to work, and for reductions in working hours.[16] However, between these two conventions the National Commission on Technology, Automation, and Economic Progress, which was to mark a turning-point in the whole debate, was established in December 1964.

The legislative mandate of the Commission was as follows:

(i) To assess the past effects and the current and prospective role of technological change;
(ii) To assess the impact on production and employment over the succeeding ten years;
(iii) To define those areas of current community and human needs towards which application of new technologies might most effectively be directed;
(iv) To assess the most effective means for channelling new technologies into promising directions, and to assess the proper relationship between government and private investment in the application of new technology;
(v) To isolate measures which local, state, and federal government might take to promote both technological change and adjustment.

(NCTAEP 1966, xiv.)

Membership of the fourteen-person Committee was, in UK terminology, tripartite and included three labour leaders.

The deliberations of the Commission are themselves of considerable interest. Using rates of increase in output per man-hour as an index of technological change, the Commission argued that there had been some acceleration of change in the post-war period, a conclusion affirmed by studies of reduced invention–innovation lags and of diffusion of specific innovations within the post-war US economy (NCTAEP 1966, 2–6). However, the essentially Keynesian stance adopted by the Commissioners in practice shifted the emphasis away from the technology itself. They argued that

> technological change is an important determinant of the precise places, industries and people affected by unemployment. But the general level of demand for goods and services is by far the most important factor determining how many are affected, how long they stay unemployed and how hard it is for new entrants into the labor market to find jobs . . . .
>
> It is the continuous obligation of economic policy to match increases in productive potential with increases in purchasing power and demand. (NCTAEP 1966, 9.)

This approach, despite a note of dissent from the labour representatives, encouraged the view that a future rise in unemployment would be the fault of public policy, not technological change;[17] the Commission was encouraged by the fall in unemployment from 5.1 to 4 per cent during its own lifetime, during which time 'adequate fiscal policies' (in the Commission's view) were pursued (NCTAEP 1966, 10, 27).

Despite the view of collective bargaining quoted as the epigraph to this volume, the Commission could reach no agreement on recommendations about negotiated reductions in working time. It covered all of the issues in its mandate, but its report is studded with qualifications and further notes of dissent from the labour members.[18] Although it studied reports on technological change in particular industries, no specific technologies are discussed in the report itself. The recommendations, published in 1966, included the following suggestions: that the government be the employer of 'last resort' for hard-core unemployed; that a system of family income maintenance be established; that compensatory education be given to those from disadvantaged environments; that a national job-matching system be established; and that employment assistance be delivered on a federal rather than state basis and receive adequate resources.[19]

Few of the Commission's recommendations were acted upon. Two years later, Mansfield was to describe its report as a 'bold program which thus far has received no strong or widespread support' (1968, 150). In fact, the political stage of the late 1960s was dominated rather more by domestic, racial, and overseas military considerations than by new technology (Blitz 1969, 96–7). However, one remarkable feature of events after the publication of the report was that automation practically disappeared as a source of concern at AFL–CIO conventions. The 1967 Convention scarcely discussed the report and carried a wholly uninventive motion on automation, all the demands of which had been seen before.[20] In 1969, these elements were incorporated into a policy document, *Adjusting to Automation*, which called for contract provisions on advance notification, intra- and inter-plant transfers, retraining, pay protection, early retirement, continuation of insurance coverage, renegotiation of job classifications, and gain sharing (AFL–CIO 1969): all such provisions existed in contract language negotiated by affiliates at the time (Blitz 1969). However, the biennial conventions of the period 1969–79, in a rather remarkable parallel to Trade Union Congresses in the UK, contain virtually no discussion of automation and technological change.[21]

The similarity of events breaks down in the absence of a second automation debate within the AFL–CIO. Concern with microprocessor-based innovation emerged again in 1981 in a rather diluted form with a motion from the Department for Professional Employees (the white-collar section) recognizing 'the broad nature of technological advance' and calling upon the Federation merely to 'expand its capabilities' in the area; the motion was referred.[22] Whereas in the UK there was much debate in the late 1970s about the possible disastrous effects of new technology, in the USA there was a concern with the slow-down in productivity growth (see Chapter 3).[23]

Nevertheless, it does seem possible to explain AFL–CIO policy development in similar terms. Two factors stand out in an explanation of the demise of the first automation debate. The first is economic: the period of 1961–9 was one of steeply rising manufacturing output *and* employment in the USA (see Figure 2.3). Indeed, the period 1961–79 did not see the reappearance of the problems the Commission saw in the 1950s: whereas in 1961 manufacturing employment was lower and output higher than in 1951, the three brief falls in employment between 1969 and 1979 were accompanied by falls in output, were not noticeably the consequences of displacement through innovation, and in any event did not significantly interrupt the trend rise.

As a result, the nature of the anxieties confronting the AFL–CIO changed somewhat. Employment projections produced by the

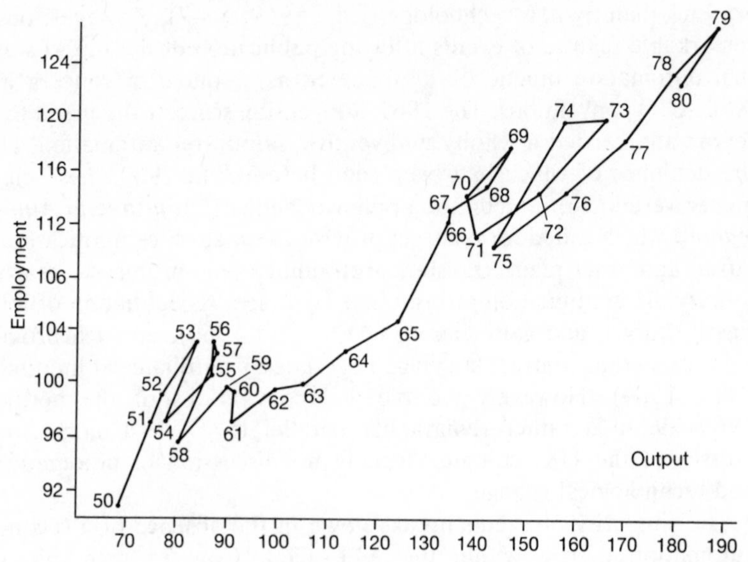

Figure 2.3. Manufacturing Output and Employment, USA 1950–1980 (1962 = 100).
Source: Freeman *et al.* 1982, 155

Bureau of Labor Statistics in 1981 indicated the possibility of short-term displacement of labour through technological change in the period to 1990, but in the context of substantial overall employment growth: the projections showed variations from 23 to 31 per cent growth over 1978 employment levels, with unemployment varying from 4 to 6 per cent (GAO 1982, 11). Many of the growth areas identified were, of course, in areas where AFL–CIO affiliates had not succeeded in recruiting, but the growth of employment in non-union sectors presents substantially different problems from an overall decline in employment. Since, according to several analysts of the US post-war scene, the level of economic activity has been an important factor influencing US labour union responses to change (Levinson *et al.* 1971; McLaughlin and Miller 1979), policy development at that time is likely to have been influenced by such generally favourable projections.

The second factor is the notable success achieved by labour unions in the 1950s and 1960s in securing contract provisions covering technological change. In 1959, pathbreaking agreements at the Kaiser Steel Corporation and the Armor Meat Company established job-security provisions and long-term adjustment strategies. The Pacific Coast Mechanisation and Modernisation Agreements of 1961 and 1966, and the agreement between the International Association of Machinists and first Boeing in 1963 and secondly Lockheed in 1965, were similarly major breakthroughs in the negotiated adjustment to change. In the late 1960s the AFL–CIO, which had historically very little involvement in the collective bargaining activities of its affiliates, tended simply to rubber-stamp the achievements of these affiliates at company and industry level. The centre of attention at the time was local or industrial rather than national (Schultz and Weber 1966; Blitz 1969; Levinson *et al*. 1971).

In the 1970s notable extensions to the coverage of provisions providing for security in the face of technological change occurred in the newspaper industry, in food retailing, and, latterly, in the vehicle industry (Murphy 1981; Wheeler and Weikle 1983; Willman 1983b). As a result, few highly unionized industries faced job loss through technological change without some form of contractual protection.

The results of BLS surveys of thirty industries affected by technological change in the 1960s and 1970s are presented in Table 2.2. No industry experienced output contraction, less than one-third experienced employment contraction, and in only five—footwear, bakery products, aircraft, refining and public transportation—is there sizeable union membership unprotected by contractual provision. As a consequence, whereas in the UK the TUC felt able to offer collective bargaining advice to affiliates confronted by new technology, activity in the USA was never co-ordinated through the AFL–CIO: the very success of affiliates themselves discouraged it.

However, this begs the question of the relationship between the confederations, their affiliates, and government which is important in understanding further differences between the two countries. Several authors have remarked that the AFL–CIO was amongst the weakest of national confederations in the 1970s, both in terms of its ability to influence governments and in terms of its control over affiliates (Windmuller 1975; Clegg 1976): more specifically, the Confederation tended to have almost no influence over the decentralized collective bargaining behaviour of its affiliates, and was unable to exert influence over a wide range of issues within the American political system (Thomson 1981, 336–41). The absence of a long-term close and stable link with a particular political party and the fact that

Table 2.2. Union Membership, Contract Provisions, Employment, and Output for Thirty US Industries, 1960–1980

Industry	Period	Union coverage[a] %	Provisions on technical change[b]	Employment trend	Output trend
Textiles	1960–74	15	No	+	+
Lumber and wood	"	41	Yes	+	+
Tyres and tubes	"	90*	No	+	+
Aluminium	"	n/a	No	+	+
Banking	"	4	No	+	+
Health services	"	n/a	No	+	+
Apparel	"	56	Yes	+	+
Footwear	"	50*	No	–	+
Motor vehicles	"	66*	Yes	+	+
Railroads	"	100	Yes	–	+
Retailing	"	25	Yes	+	+
Bakery products	1960–77	80	No	–	+
Concrete	"	50	No	+	+
Air Transport	"	n/a	Yes	+	+

Telephone communications	"	70*	Yes	+	+
Insurance	"	4	No	+	+
Pulp and paper	"	90*	Yes	−	+
Cement	"	90*	Yes	−	+
Steel	"	95*	Yes	−	+
Aircraft	"	n/a	No	−	+
Wholesale Trade	"	25	No	+	+
Coal	"	66	Yes	+	+
Oil and gas	"	40*	No	+	+
Refining	"	90	No	−	+
Public transportation	"	n/a	No	−	+
Electric & Gas utilities	"	50	Yes	+	+
Meat	1960–80	33*	Yes	+	+
Foundries	"	60*	No	+	+
Metalworking	"	33	No	+	+
Electrical/electronics	"	34	No	+	+

[a] At end of specified period; * refers to collective bargaining coverage; n/a, not available.
[b] 'Yes' indicates more than 20 per cent of employees covered.

Source: BLS, Bulletins 1817, 1856, 1961, 2005, 2033, 2104.

during much of the post-war period some of the largest US unions remained outside the AFL–CIO may have been important here (Clegg 1976, 52).

In any event, no developments parallel to the close relationship between confederation and party characteristic of the UK in the 1970s occurred, nor did any parallel co-ordination of bargaining activity. Nevertheless, one interesting feature does emerge from the international parallels drawn here. Once trade union policy on change has developed away from localized collective bargaining adjustment to the suggestion that thoroughgoing governmental action—primarily involving the maintenance of full employment—is necessary to cope with technological change, the policy has reached the end of the line: either a broadly sympathetic government follows the appropriate policy, in which case technological change may recede as a problem, or it does not, in which case the policy is frustrated. In both cases the trade union movement does not rely on factors within its own control. The former case describes the experience of the AFL–CIO in the late 1960s, the latter that of the TUC in the period since 1979.

The emergence of a second automation debate within the AFL–CIO was thus hampered by the feeling of *déjà vu* and the absence of new policy development after 1967. This in turn raises the question of why such new developments failed to appear: different economic circumstances are only part of the answer. I have already argued that the development of new initiatives between the two UK debates was influenced by the growth of white-collar unionism. It is thus of some interest to note that the attempt to revive the debate within the AFL–CIO in 1981 came from a numerically weaker white-collar section, the DPE.

In 1978, approximately 18.7 per cent of union membership affiliated to the AFL–CIO was classified as white collar. This percentage had grown from 12.2 per cent in 1960 and was concentrated quite heavily in public-sector employment: white-collar private-sector industries, such as banking, and white-collar employees in manufacturing, such as foremen and technicians, remain largely unorganized (BLS 1980). Since in both countries white-collar union organizations have led in the promotion of initiatives on microprocessor technology, the relative weakness of white-collar unionism in the USA may thus be one factor underlying the truncation of the second automation debate.

Significantly, the qualitative developments characteristic of UK concerns with the design stage have not appeared: a survey of collective agreements conducted by the DPE found that, although concern with issues related to technological change was more pervasive than previously was the case,

> These issues are similar if not identical to those found by others who have studied collective bargaining responses to technological change as far back as 1965. In our inspection of current contracts, there was no new type of clause or provision that was not discussed in these earlier studies. Recent contracts may demonstrate a greater pervasiveness of technological change provisions; or a greater sophistication or favorableness of language, but the basic strategies and types of clauses remain the same. (Murphy 1981, 29).

This conclusion is confirmed in the survey work of McLaughlin and Miller (1979) and in the publications both of government departments (BNA 1979; BLS 1980) and of the Industrial Union Department of the AFL–CIO (AFL–CIO 1980): the coverage of protective provisions may have expanded, but they did not apparently change in nature. Following the foregoing discussion, this too may be the result of the relatively low levels of union organization among white-collar occupations able to make the appropriate input.

IMPLICATIONS

On the basis of a wider but less detailed analysis of trade union responses which also includes Australia and parts of Europe, Corina identifies a number of common prescriptive components of 'the trade union view of new technology'. These include a fear of job loss, both through rapid innovation *and* through a failure to innovate, a desire to participate in decisions about change, and, above all, a reliance upon government action to promote innovation and to prevent unemployment (Corina 1983, 180–3). The policies of the AFL–CIO and TUC have thus been broadly typical.

The pattern of a growing concern with governmental action in the area is not particularly surprising. Most union movements engage in some kind of political activity. Indeed, Crouch has argued that such a concern is an inevitable accompaniment of their dual concern with labour costs and unemployment (Crouch 1982, 202–6). There appears to be no good reason why policies on technological change should be insulated from this general trend. However, in the UK and

USA, recourse to government action has failed to shift the burden of adjustment to technological change in the presence of trade unions away from a reliance on collective bargaining. This raises at least two sets of questions: the first concerns the success of the bargaining mechanism, while the second concerns the contrast with other situations where government action is important. For unions which see the need both to have and to control innovation there also emerges the need to impose adjustment costs without generating a disincentive to invest. The paradox, noted also by Corina, is that

> mobilised workgroups can provide the foundation for securing the right to a negotiated share of technological change benefits, but they can also impede productivity gains through technological change in the first instance. (1983, 187.)

This entails a rather different view of the relationship between trade unions and technological change from that outlined in Chapter 1. The collective bargaining tactics of trade unions tend to focus on the impact of technological change upon the job-holder, requiring wider government policies to lubricate the process by expansion of employment. In the absence of such policies, union strategies may disintegrate into mere resistance, which ultimately proves self-defeating as the failure to innovate causes firms or industries to succumb to competitive pressures from those companies (or economies) which have not experienced similar resistance.

As Crouch has shown, this conclusion requires no theory of shop-floor militancy: it simply follows from the logic of collective action. The imposition of high adjustment costs by a small group has only a small negative consequence for the economy as a whole, but may result in considerable short-term gains for the group itself; conversely, acceptance of change at low cost may mean comparative disadvantages for the group in the short term. If each group follows the logic of its interests, imposing costs on changes which are advantageous to it and resisting those which are not, collective benefits are ultimately forgone (Crouch 1982, 198–200). The logic applies particularly to the extremely decentralized bargaining systems of the UK and USA compared with those in parts of Europe: in fact, decentralization of bargaining structure is an important part of Thomson's explanation for the increasingly marginal role played by the American industrial relations system in wider movements for social change (1981, 304f.).

From this perspective, Corina's paradox is resolved in terms of a theory of interest-group representation. In the short run, workers will resist change because it affects job security and requires sometimes inconvenient accommodation in the form of retraining, mobility, altered shift patterns, and so on. However, in the longer term, the more adaptive, more highly productive economies or companies have relatively more jobs to offer. A configuration of interest groups which prevents sectional behaviour through the ability to promote global policies of accommodation to change is thus, *ceteris paribus*, intrinsically more efficient than one which relies upon sectional bargaining behaviour. The argument here is straightforwardly drawn from that of Olsen who argues that

Distributional coalitions slow down a society's capacity to adopt new technologies and to reallocate resources in response to changing conditions, and thereby reduce the rate of economic growth. (1982, 74.)

The contrast between these two configurations has been discussed in some detail by writers who seek to contrast the UK with West Germany. Hotz, for example, suggests that local bargaining over the terms of change relying on decentralized union power in the UK can be unfavourably compared with the greater ability of a technically more competent German management to effect rapid change. He concludes that institutional arrangements for the resolution of employer–employee conflict in the UK foster 'mistrust and scepticism between management and workforce, and a snowball effect can lead finally to low investments, low wages and little improvement in productivity' (Hotz 1982, 351). Streeck presents a more general argument in comparison of German and British car manufacture which relies on arguments about corporatism:

Ceteris paribus, economies with a corporate system of industrial relations would be expected to perform better than economies with a pluralist industrial relations system [because] corporatist systems of industrial relations exclude from articulation sectional interests that stand to profit from preservation of the status quo in spite of resulting sub-optimal performance of the industry or economy. (1984, 140–1.)

The contrast here is between, on the one hand, a co-operative approach to change which involves a global safety net for those adversely affected but which does not necessarily lead to efficiency losses in the obstruction of managerial intentions and, on the other, a reliance on local sectional bargaining power which does constrain managerial behaviour. The former approach implies involvement of workers' interest groups in managerial decision making as well as the involvement of government in the provision of the safety net itself. The latter approach implies an adversarial relationship between company and union where both such elements are absent.

The 'German Comparison' may be a rather hackneyed one by now, but discussions of interest-group organizations do serve to highlight the essentially economic dilemma which has confronted the TUC in the UK debates outlined earlier. So long as economic circumstances allowed the belief that technological change was ultimately benign in securing productivity improvements, concern was with the development of local adjustment mechanisms. Once the productivity–employment relationship broke down, however, efforts switched to the attempt to change economic circumstances through encouragement of government action.

The American experience is different again. Writers such as Thomson (1981, 297–343) have contrasted the role of collective bargaining in Western European economies as an agent of social reform with the more narrowly economic and localized functions it shows in the USA. Yet, within these limitations, it is possible to suggest on the basis of the foregoing discussion that this narrower bargaining function has been extremely successful at accommodating certain types of technological change. Moreover, certain American writers (e.g. Ulman 1968) are prepared to argue that the collective bargaining system in the USA characterized as 'connective' is more efficient than the more centralized 'competitive' system characteristic of the UK and, to a lesser extent, other European countries.[24] This is a very different form of argument to that of Hotz and Streeck: whereas they see corporatist arrangements which avoid sectional activity as conducive to the more efficient accommodation of change primarily because of the dilution of sectional interest by more general corporatist arrangements, Ulman emphasizes flexibility and the localized connection of productivity-determining and wage-setting activities without government involvement.

These quite general discussions are probably not wholly satisfactory for resolving the issue of what kinds of mechanisms deal best with technical change. Many would argue that the so-called 'corporatist' arrangements of West Germany—with industrial unionism and co-determination—are better at accommodating change than the decentralized collective bargaining of the UK. However, decentralized collective bargaining in the USA appears from the discussion so far to have been relatively successful compared with similar bargaining in the UK. This argument will be assessed more fully in Chapters 6 and 7.

CONCLUSION

There are at least two aspects of trade union policies towards technical change. One seeks to promote investment in new equipment, the other to control the possible negative consequences of change for unionized job-holders. In both the UK and the USA the balance between the two has been heavily dependent on the rate of technical change and on the economic climate; as unions' concern with both grows, an appeal for government economic action becomes much more likely.

However, the appeal to government by a central union confederation also depends on the success of collective bargaining mechanisms in accommodating change, and here there appears at first sight to be a difference between the UK and the USA in the success of AFL–CIO affiliates. This is of particular interest since it implies that, even between decentralized industrial relations systems not characterized by high levels of government involvement, different forms of collective bargaining may be differentially efficient in accommodating technical change.

However, it may be the case that one needs further to discriminate and suggest that certain types of change may be more easily accommodated by certain types of collective bargaining institution. The argument that the UK suffers disproportionately from trade union resistance to technical change might thus be more accurately phrased as a question concerning goodness of fit between specific institutions and specific changes, and the extent to which the UK suffers disproportionately from experience of institutions which cannot accommodate the current generation of changes. To assess this argument one needs some grasp of the extent to which resistance is experienced, a typology of the sorts of changes occurring, and a theoretical approach to institutional change: these three topics will be the subject of Chapters 3–5.

NOTES

1. In the early 1960s the Women's TUC and the Non-Manual Workers Advisory Committee produced a report on the impact of office automation which suggested that 'few redundancy problems had so far confronted non-manual unions'. (TUC *Report* 1963, 129).
2. TUC *Report* 1957, 249, Agenda motion 50; *Report* 1963, 381.
3. Composite 4, 'Automation, Technology and Microelectronics'. As noted in this chapter (in 'Continuity and change'), it is of some interest that all the supporters of this motion—TASS, NUBE, SCPS, APEX, and ASTMS—are white-collar unions (*Report* 1978, 477).
4. Most of the representatives in the Employment and Technology Group were research staff rather than 'front-line' officials or general secretaries. In addition, the major print unions—NUJ, NGA, and SOGAT—who were at that time resisting the impact of new technology (see Chapter 6) did not attend, preferring to develop rather different policies and to co-ordinate their activities through the more permanent Printing Industries Committee. The Report was accepted by Composite 20, 'New Technology' moved by the AUEW(E) and seconded by the EETPU (*Report* 1979, 642–3).
5. For summaries of such developments, see Robins and Webster (1982), Manwaring (1981)). In 1980–1, the TUC alone mounted 114 New Technology courses of not less than 2 days' duration (*Report* 1981, 191).
6. Composite 13, 'Economic Policy and Unemployment', called amongst other things for control of technology; it was moved by the TGWU and seconded by the AUEW(E). Composite 17 focused on 'New Technology, Job Opportunities and the 35-hour Week'; it was moved by USDAW and seconded by the NGA (*Report* 1980, 588–90).
7. The activities reported primarily related to the Public Services Committee and the Social Insurance Department of the TUC itself (*Report* 1981, 129).
8. The object of this service was to 'minimise mismatches in the labour market' (*Interim Report*, 49).
9. For a discussion of innovation diffusion, see Chapter 3.
10. By 1975 there were industry committees in steel, construction, local government, transport, health services, fuel and power, textiles, clothing, hotels and catering, and printing.
11. AFL–CIO Constitutional Convention *Proceedings*, 1955, 193–4.
12. *Proceedings* 1975, resolution 140, 357 f. The resolution covered the content of other motions submitted by the woodworkers' and railroad unions (Nos. 36, 61).
13. *Proceedings* 1959, resolution 115.
14. *Proceedings* 1961, resolution 198, quote from VI, 436.
15. *Proceedings* 1963, resolution 222, VI, 417–18.

16. *Proceedings* 1965, resolution 197, VI, 384–7.
17. The view of the three labour representatives, Beirne (Communication Workers), Hayes (Machinists), and Reuther (UAW), was that the overall report 'lacks the tone of urgency which we believe its subject matter requires' (NCTAEP 1966, 6).
18. See, for example, NCTAEP (1966), 6, 41, 42, 56, 58, 93, 107.
19. For the full list, see NCTAEP (1966), 110–12.
20. *Proceedings* 1967, resolution 218, 435–6. This resolution focuses on economic expansion, collective bargaining adjustment, retraining, reduced working hours, income maintenance, and the establishment of a technology clearing-house.
21. The exceptions were: resolution 84 of the 1969 Convention calling for full crews on Amtrack (*Proceedings* 1969, 80); and resolutions 92 in 1971 and 157 in 1973 calling for reductions in the working week (*Proceedings* 1971, 142–3; 1973, 411); the latter motion was referred.
22. *Proceedings* 1981, resolution 165, 249.
23. The contrast here is made by Wheeler and Weikle (1983). They note the concern of the Kennedy administration with encouraging automation and full employment, whereas 'in more recent times, the code word for concern over mechanisation has been "productivity"' (16). They also contrast Reagan's 'National Productivity Advisory Committee', primarily concerned to raise productivity indices, with the earlier role of the NCTAEP.
24. As Thomson notes (1981, 331), although he discussed efficiency, Ulman was mainly concerned with the inflationary consequences of different structures.

3

The Evidence of Trade Union Resistance to Change

INTRODUCTION

BECAUSE trade union resistance to change may take several forms, different types of evidence need to be considered at this point. If the threat of trade union opposition or resistance causes employers to shy away from innovation in the UK, one might expect this to show up in the form of low relative notes of innovation and relatively low levels of capital investment. The first set of relevant data thus concerns rates of innovation. Logically, the second set of data concerns the adoption of innovations, and one thus needs to look at diffusion rates within the UK.

Since such evidence says nothing about trade union behaviour *per se*, another important set of data relates innovation to strike activity. One might expect under the hypothesis of generalized trade union resistance that more innovative industries would experience a larger number of strikes associated with resistance to change and, more specifically, that one might identify instances of conflict over technical change.

However, another form of the resistance hypothesis suggests that strike activity might not be the best evidence: rather one should look for restrictive practices and feather-bedding. Two further forms of evidence are relevant here: the first consists of the demonstration that part of the UK 'productivity problem' is attributable to the activities of trade unions in the face of technical change; the second consists of direct evidence on the retention of restrictive practices after change. I shall look at these five different types of evidence in turn.

INNOVATION IN THE UK

A great deal of evidence of several different types supports the view that the UK experiences some form of 'innovation problem', although the data do not lead to precise definition of its nature. The evidence is based on various measures of innovation, including patent statistics, R&D expenditure, and the size and age of capital stock.

Pavitt and Soete approach the problem using data on patents achieved in the USA by foreign countries as an index of their respective levels of innovative activity. They calculate that 'British innovative activities have declined noticeably compared with those of other countries over the past 10 to 15 years, indeed over the past 70 years' (1980, 55). They are able to demonstrate the existence of a strong relationship between exports per caput and US patents per caput at industry level; the UK, with a relatively low share of US patents, fares badly in export performance outside the aircraft and defence-related sectors.[1]

Other work on patenting activity confirms this conclusion, differing primarily in terms of the time-scale over which the failure to innovate is identified. Freeman *et al*. characterize the period since *1883* in the UK in terms of

> a low ratio of domestic investment with, as a consequence, a declining rate of growth of productivity, slow growth of exports with rapid growth of imports of manufacturers, and slow adoption of new technologies relative to competitor countries. (1982, 170.)

They note the decline of UK patents in the USA from 34.6 per cent of total foreign patents in 1883 to 10.1 per cent in 1979.[2]

If R&D expenditure rather than patenting activity is used as a measure, a similarly depressing picture emerges. Studies based on R&D expenditure have also identified a 'technology gap' relating technological superiority to export performance: the UK does not perform well (Stoneman 1983, 254–6). Moreover, the UK has fallen down the 'international R&D' table since the 1950s, and appears to concentrate R&D expenditure in aircraft and military applications. Compared with West Germany and Japan, industries such as chemicals, electrical machinery, and instruments take a low percentage share of industrial R&D (Freeman 1979, 67–9). Freeman goes on to argue that a failure to compete with the USA in these high R&D sectors exacerbated problems in 'staple' industries such as mechanical engineering, where technical effort has steadily declined. However, it is worth noting that, as late as 1979, R&D expenditure as a percentage of gross domestic product in the UK was above the OECD average, and bettered only by the USA and West Germany (Giersch and Wolter 1983, 41).

Other work by Panic and Rajan (1971) identifies slower product innovation in UK manufacturing, particularly in those sectors where patterns of demand are changing rapidly. Panic identifies the better quality, higher technical performance, and better design of foreign products, rather than their price, as factors contributing to the rapid growth of manufactured imports into the UK (Panic 1975; see also NEDO 1977, Rothwell 1979, and the industry cases reported in Pavitt 1980). At industry level, Prais (1981), comparing the UK, USA, and West Germany, identifies technological backwardness in food, furniture, and machine tools; he shows that a failure of product innovation has contributed to the decline of the UK tyre industry and the virtual disappearance of typewriter manufacture.

One might expect that this failure to innovate might leave the UK with a smaller, and older, capital stock than its main competitors. In fact, Freeman *et al*. show a low rate of growth of total, non-residential fixed capital stock in the UK since 1870, compared with our main industrial competitors, and since 1955 lower rates of return on capital and lower capital productivity (1982, 149–52).[3] However, in metal-working industries at least, it does not seem to be the case that the UK operates with substantially older equipment than the USA (Bacon and Eltis 1974; Ayres and Miller 1981). More recently, Smith *et al*. have demonstrated that, in nine of twelve intersectoral productivity comparisons between the UK, USA, and West Germany, British industry is relatively under-capitalized, and that this in part explains the failure to attain high levels of labour productivity (1982, 47, 96).[4]

Surprisingly, they find that capital intensity per man is higher in UK manufacturing than in the USA,[5] but in none of the 117 manufacturing industries covered is British labour productivity higher than that in the USA (1982, 76). A rather different picture is presented by Prais, who distinguishes between capital used per employee and per unit of output. On the basis of his ten-industry analysis, he argues that

> Capital investment *per employee* is lower in Britain than in Germany or in the United States, but investment *per unit of output* is fairly similar in all three countries. (1981, 169.)

and

> It is as if the average British employee operates one machine producing 100

units per day, while the average American employee operates two machines each having an output of 150 units per day. (1981, 269.)

This picture is confirmed by more recent data. Labour costs in the period 1971–83 were much higher in the USA than the UK, but US unit labour costs were consistently lower than those in the UK (Ray 1984a, 64).

The picture painted here is of a combination of under-investment and overmanning, with perhaps the implication that the two go together. In fact, there is a certain logic for this, as I shall show below.

The evidence presented so far illustrates a failure to innovate, but provides no basis for attributing this failure to actual or expected costs arising from trade union resistance to change. In fact, the evidence lends equal support to trade union arguments about managerial failures to innovate. However, evidence linking failure to innovate with union organization does exist in the R&D literature. It is based on American research into the impact of R&D expenditure at industry level which includes union density as a control variable. Estimated union effects are consistently and significantly negative; as Addison notes, the general conclusion is that 'more heavily unionised firms or industries grow at a slower rate than their less organised counterparts' (1983, 295). One is left to wonder whether slow growth follows from unionization, or vice versa.

This work has not been replicated in the UK and its relevance may be uncertain. However, more direct evidence of trade union resistance in the UK is available. Prais, for example, found that the use of more productive equipment in the tobacco and beer industries in the 1970s was held up by problems of achieving agreement with trade unions on manning levels; low productivity in the tyre industry was seen as due to a combination of single-firm dominance and 'the power of unions to resist labour shedding in the face of technological advance and declining demand'. The British can industry in the 1970s 'was particularly hampered by labour troubles impeding the introduction of the two-piece can' (Prais 1981, 104–5, 120–1, 217–18, 257). However, one may doubt that this is evidence of a general trend. A recent PSI survey has found that only 9 per cent of new-process users and 2 per cent of new-product users encountered shop-floor resistance; however (as Prais also implies), frequency of resistance was related to plant size. Moreover, it was more common for process users in mechanical engineering (22 per cent), vehicles (12 per cent), and textiles (11 per cent) to experience trade union obstruction

(Northcott *et al*. 1982, 44–5). It thus seems likely that experience or expectation of labour resistance to change may be prevalent, at least within certain sectors: mature manufacturing industries appear particularly susceptible.

In summary, then, the innovation literature suggests that the UK suffers from slow rates of innovation and that, in some sectors, this may be related to trade union resistance. However, researchers in this field do not focus on trade union resistance as a primary or even significant variable. The two principal analytical models are in fact Schumpeterian (or 'technology-push'), in which the motor of innovation is seen to be R&D activity, or 'market-pull', where demand is the primary agent of change. Researchers concerned to assess their relative usefulness do not normally concern themselves with measurements of trade union resistance or bargaining power, or even with labour costs (Freeman *et al*. 1982, 35–44; Kamien and Schwartz 1982, 31–48). In addition, the literature contains no discussion of the mechanisms through which resistance might occur. Freeman *et al*.'s result, for example, suggests a period of comparative innovative disadvantage almost as long as the gestation period of the UK trade union movement itself, during which many of the features of trade union activity alleged to inhibit productivity and growth have appeared. If there is a relationship between innovation and trade union behaviour, one is led to speculate about its direction.

INNOVATION DIFFUSION

Similarly, the literature on the spread of innovations pays little attention to the role of trade union activity. Much of the theoretical work in the area consists of modelling the diffusion process: the most commonly observed pattern is the S-shaped logistic epidemic curve depicting rates of adoption within a given industry or across a sample of firms (Davies 1979, 8–36; Stoneman 1983, 65–151). Empirical work consists either of the study of the diffusion of particular innovations (Nabseth and Ray 1974; Ray 1969, 1984b) or general national studies of several such changes (Davies 1979).

Within this framework, trade union resistance might affect the overall rate of innovation diffusion within an economy (i.e. the slope of the logistic curve) or the sequence of adoption (i.e. the position of unionized firms or industries *on* the curve). Davies' systematic attempt to assess the factors affecting diffusion speed suggests that both the profitability of innovation and industrial structure are of primary importance in encouraging adoption. However, labour-intensive industries also tend to be rapid adopters, which Davies interprets as supporting Salter's view that the age of capital stock in labour-intensive industries may be a factor (Salter 1960; Davies 1979, 141–2). The latter finding is also consistent with hypotheses relating innovation decisions to the requirement to reduce labour costs where these are an important element in total costs.

Studies of the diffusion of particular innovations do not, with one exception, show the UK at a particular union-related disadvantage. Stoneman's study of computer diffusion and Fleck's of robotics uncover little or no evidence of labour resistance: in the former case, trade union activity appears neither as a reason for adoption of computers nor as a reason for their rejection (Stoneman 1983, 139; Fleck 1982). Of the innovations studied internationally by Nabseth and Ray, only in the case of the diffusion of computer numerically controlled machine tools does the UK suffer a disadvantage.

This case is of some interest, and Nabseth and Ray's attempt to 'score' the factors affecting diffusion appears here as Table 3.1. It can be seen that the union-related disadvantage suffered by the UK is outweighed by that of the USA, but that the wage effect more than offsets it. The case raises some important issues about the mechanisms of trade union resistance to change. If the picture applies more generally, it may be the case that the combination of factor prices and trade union attitudes does serve to slow down the rate of innovation and diffusion in the UK. In the USA, labour is relatively expensive, and the desire arises to substitute capital for it; occasionally, this desire is shared by union leaders who promote high-productivity high-wage policies (Mansfield 1968, 150–1). In the UK, labour is comparatively cheap and there is less incentive to substitute capital but, so the argument runs, it is intractable, and the benefits of capital investment can be dissipated by the maintenance of restrictive practices.

I shall have more to say about the latter below, but this argument does serve to illustrate further problems involving the nature of trade union resistance. A priori, exacting a price for technical change should *accelerate* innovation by changing relative factor prices; it requires the additional argument that the gains from change will be dissipated by restrictive practices to retain the conclusion that trade union activity serves to hamper innovation. Bell (1983), for example, has suggested that *incremental* rather than radical change might be

Table 3.1. Factors Affecting the Diffusion of Numerically Controlled Machines

		Austria		Italy		Sweden		UK		USA		West Germany	
	Weight	Value	Score	Value	Score	Value	Score	Value	Score	Value	Score	Value	Score
Wage level	40	+1	40	—	—	+3	120	+2	80	+5	200	+2	80
Importance of aerospace industry	10	—	—	+1	10	+1	10	+4	40	+5	50	+1	10
Quality of information system	10	+2	20	—	—	+4	40	+4	40	+5	50	+2	20
Investment financing possibilities	10	—	—	—	—	+4	40	+3	30	+5	50	+1	10
Management attitudes	5	—	—	—	—	+3	15	+2	10	+5	25	+1	5
Condition of the market	5	+1	5	+1	5	+3	15	+4	20	+3	15	+5	25
Trade union attitudes	5	—	—	—	—	—	—	−2	−10	−4	−20	+1	5
Technical factors	5	—	—	+1	5	+2	10	+3	15	+5	25	+1	5
Labour-market conditions	5	—	—	+1	5	+5	25	+2	10	—	—	+4	20
Other relevant factors	5	+2	10	+3	15	+3	15	+5	25	+5	25	+5	25
Total	100		75		40		290		260		420		205

Source: Nabseth and Ray 1974, 56.

particularly vulnerable to this kind of union behaviour. Where employers wish to introduce small changes which generate minor efficiency improvements, trade union resistance may raise costs above benefits: over a period of time, he argues, substantial efficiency losses may occur.

However, other evidence points in a rather different direction. For example, it appears that expectation of bargaining over change may 'shock' companies into preparing the case for change more thoroughly, and thus developing both watertight arguments and alternative implementation strategies (Wilson *et al*. 1982). Expectation of a strike may simply encourage companies to cost change in a slightly different way, so that

> Against the probable gain in market share due to cost reduction must be set the probable loss of market share (due to loss of custom) weighted, in turn, by the probability of the occurrence and duration of a strike. (Ulman 1968, 336.)

Thus, perhaps the most severe threat to the returns on innovation concerns the retention of outmoded practices or manning levels which could pose a recurrent drain on such returns. Ulman goes on to note that a peculiarity of the UK bargaining system is the range of restrictive practices which have, in his view, 'tended to depress capital–labor and output–capital ratios and also, thereby, investments' (1968, 360). Subsequent evidence collected by Pratten illustrates how damaging this can be. He notes that labour productivity and earnings per hour in UK manufacturing industries were generally lower than in their counterparts in the USA, West Germany, and France.[6] His case-studies show that the impact of the combination of the restrictive practices and low pay may hold back productivity *and* innovation. Of the comparisons between the UK and the USA, 22 per cent revealed that restrictive practices depressed labour productivity in the UK; however, a further 22 per cent argued that higher US wages pressured companies operating there to automate operations or to invest in new equipment (Pratten 1976, 81–105).

Whether one is concerned with the first use of a new technique or the adoption of methods diffusing through an industry, it is clear that, at least within manufacturing, UK firms generally face the tasks of consultation or negotiation with trade unions. There is substantial evidence that industrial relations considerations tend not to be central to decisions about production methods or capital investment decisions (Piore 1968; Batstone 1984, 42–8). However, where modifications to working practices are involved, companies find themselves in negotiation over the *consequences* of change. Batstone compares the results of four surveys—in 1972, 1976, 1980, and 1983—which appear to suggest that the likelihood of negotiation over changes to working practices increased substantially over the period: 89 per cent of unionized firms in his own 1983 sample would enter into such negotiations. He concludes that, 'while it should be recognised that the extent to which managerial prerogative was "invaded" is very limited, it is nevertheless the case that a wider range of issues became the subject of bargaining' (Batstone 1984, 131–6, 242–3).

Although one is not exactly comparing like with like, it is of interest to quote evidence from the USA over the same period. McKersie and Klein quote a range of survey evidence showing both a decline in the extent of trade union obstruction of change and a reduction in 'feather-bedding' in the USA during the 1970s (1982, 5, 13). Slichter *et al*. (1960) had previously referred to a similar decline in the 1950s, arguing that the 'work rule' problem was a feature of wartime labour controls which had been successfully eradicated in subsequent negotiations. Both sets of evidence support McKersie and Klein's conclusion that 'a key strength of the US system of collective bargaining has been its ability to grapple with productivity problems' (1982, 76). This suggests that a problem for the UK might be the inability to shed work rules during periods of bargaining over change. The existence of negotiation over change or some initial resistance to it does not necessarily imply that there will be some long-term productivity loss: this will only occur if the *outcome* of the bargaining process increases manning arrangements or output controls which are inefficient. I shall return to this point in the section on 'Inefficient labour utitilization'.

These considerations reinforce the suggestion that there are at least two sources of risk for potential innovators which relate to trade union activity. The first is the prospect of a strike of indeterminate length. Batstone, using evidence from the Warwick survey (Brown 1981), indicates that industrial relations considerations are more prominent in innovation decisions in strike-prone establishments (1984, 46–7). The second source of risk is that bargaining will result in the retention of outmoded working practices on new equipment over a longer period of time. The remainder of this chapter will look at each in turn.

INNOVATION AND STRIKES

Evidence already quoted, such as that of Prais, shows that strikes may impede technological change. Moreover, there are other highly publicized instances of trade union resistance: for example, containerization disputes on the docks and resistance to photocomposition in national newspapers. However, it is extremely difficult to make an overall assessment of the extent of strike activity in opposition to technical change in the UK within, for example, the last thirty years, primarily because of the unavailability or inadequacy of data.

The first set of difficulties concerns the analysis of strike activity by *cause*. A number of authors have noted that there are relatively severe problems of interpretation and discovery (i.e. of the reasons people give for a strike) as well as of attribution of causality (i.e. the role of such reasons in a causal explanation of strike activity) (McCarthy 1959; Hyman 1972; Durcan *et al*. 1983). However, for our present purposes these problems pale beside the fact that 'technological change' has not been a principal cause in any of the official listings of principal causes since 1946.[7] In fact, resistance to technological change might be submerged under several official headings of immediate cause, such as 'redundancy' or 'demarcation', or simply under 'wage increase' if bargaining over the price of change resulted in a dispute. There is some slight evidence to suggest that redundancy and demarcation disputes have increased during periods of rapid process innovation, but this is highly impressionistic.[8]

The extent of the difficulties encountered can be assessed by analysing the list of principal stoppages given annually by the Department of Employment: the list details all stoppages within a given year which account for the loss of 5 000 or more working days, with a brief, usually one-sentence, account of the cause or objective. It represents the best source of general information about disputes over technological change but suffers from at least two disadvantages as an indicator of the level and distribution of such disputes. The first is simply a more tantalizing example of the problem discussed in the previous paragraph: some disputes clearly involve change in working arrangements, work allocation, or work-speed and may well be linked to technological change, but such change does not feature as part of the cause or objective. The second problem is that it is an inconsistently biased sample of all industrial disputes. The number of prominent stoppages and the percentage of overall working days lost for which they account vary from year to year, but in all years a disrup-

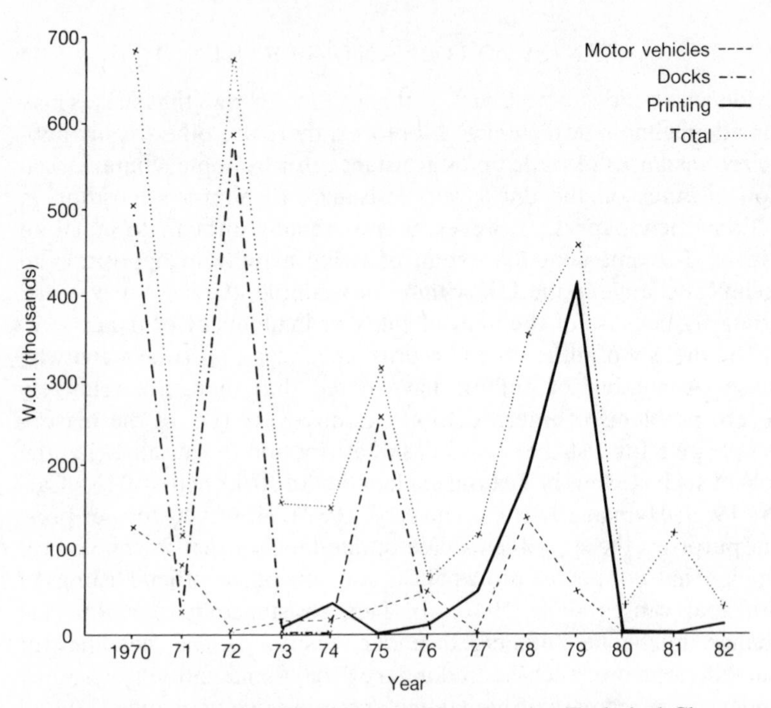

Figure 3.1. Working Days Lost in Disputes over Technical Change
Source: DE Gazette

tive pattern of small stoppages in the face of technological change or even a prolonged stoppage of a small group would be omitted.[9]

Bearing these problems in mind, Figure 3.1 extracts from the prominent stoppages since 1970 those which were explicitly to do with operation, payment, manning, or demarcation issues over new equipment. A rather marked pattern emerges. Three industries—docks, printing, and motor vehicles—account for the greater part of all working days lost (w.d.l.) through disputes over technological change, but they do so in different ways. Both docks and printing experienced 'phases' of conflict from this source over the period, in docks from 1970 to 1975 and in printing from 1977 to 1979; the motor vehicles industry displays a rather different pattern of more drawn-out recurrent lower levels of conflict. Analysis of the patterns of conflict in other industries reveals this latter pattern to be rather idiosyncratic, since the 'burst' of conflict over technological change characterizes textiles (90 800 w.d.l. between 1971 and 1974), rubber (29 800 w.d.l. between 1973 and 1974), and metal manufacture (155 000 w.d.l. between 1973 and 1975). Only the very broad category of mechanical engineering displays any similar traits, with disputes in five of the thirteen years under consideration, but this apparent recurrence is probably a function of the level of aggregation of the DE classification. Moreover, the 1970s were not untypical in displaying this pattern of concentration. If one extends the analysis back to the period 1960–9, both the concentration and the specific industries are the same: metal manufacture (179 000 w.d.l.), vehicles (126 000 w.d.l.), and dockwork (36 000 w.d.l.) are the principal culprits.

I shall return to more detailed discussion of these particular patterns of conflict in Chapters 6 and 7, but from Figure 3.1 it does seem to be the case that, over the last decade or so, most industries have been free from large disputes over technological change and that most of those which have experienced such conflict have done so over a relatively short period of time. Of course, it could be objected to this that the absence of response reflects the absence of stimulus: the absence of such conflict reflects the absence of innovation in the majority of industries.

In the absence of detailed and comprehensive information on the pattern and nature of industrial innovation, the test of such a hypothesis is extremely difficult. However, a partial test is possible on the basis of the SPRU innovation data bank covering about half (probably the most innovative half) of British manufacturing industry (Townsend *et al.* 1981; Pavitt 1983, 1984) and Durcan *et al.*'s encyclopaedic study of post-war strike activity (1983). The basic data for these MLHs where some approximate fit between the respective classifications was possible is presented in Table 3.2. As one might expect on other grounds, there is no simple statistical relationship between innovation frequency and numbers of working days lost across the post-war period;[10] other factors are far more important in the explanation of strike activity. The appropriate focus of explanatory effort is on the reasons why industries experience problems with technological change at particular points in time.

One final issue needs to be considered here: namely, the impact of a threat to strike on the adoption of an innovation. As Davies' work has shown, the profitability of change affects the rate of diffusion. This calculation will be affected if the risk of a strike must be entertained, so that change may be thwarted even where no strike occurs.

Table 3.2. Innovation and Industrial Conflict

Industry (MLH)	No. of innovations used,[a] 1945–80	Average working days lost per 1000 workers,[b] 1945–73
Food (211–29)	48	59.9
Chemicals (271–2, 275–9)	55	63.0
Iron, steel, and metal manufacture (311–23)	123	411.7
Non-electrical engineering (332, 335, 339, 341, 354)	83	374.1
Electrical engineering (364–7, 369)	135	363.8
Shipbuilding (370)	78	1388.4
Vehicles (380–1)	159	886.0
Textiles (411)	370	56.0
Paper and board (481)	33	107.1

Sources:

[a] Pavitt (1984), Table 3. The following MLH headings for which Pavitt has data were excluded because of the difficulty of matching to Durcan *et al.*'s categories: 431, 450, 463, 464, 496. In addition, several MLHs have been combined.

[b] Durcan *et al.* (1983). The average for Food (MLH 211–29) is a simple unweighted average of the categories grain milling, bread and flour, and other food. Durcan *et al.* amalgamate marine engineering into the shipbuilding category.

As Waterson (1984) has shown, there needs to be an element of monopoly for many companies to risk being first movers in the adoption of change, and change may be impeded to the extent that union threats diminish the prospect of monopoly returns. However, data which allow more systematic analysis of the effect of strike threats on innovation or adoption do not exist.

LOW PRODUCTIVITY

The existence of low labour productivity in the UK, particularly compared with the USA, has been comprehensively documented over a long period of time. Rostas (1948) shows a US/UK productivity ratio of 2.2 : 1 for the period 1935–39 on the basis of a 31-industry comparison. Paige and Bombach produce a ratio of 2.7 : 1 for manufacturing as a whole in 1950 and Prais estimates a value for the same ratio of 3 : 1 for manufacturing as a whole in 1979 (Paige and Bombach 1959; Prais 1981). Smith *et al.* show sectoral ratios for 1977 varying from 3.42 : 1 in extractive industries to 1.7 : 1 in construc-

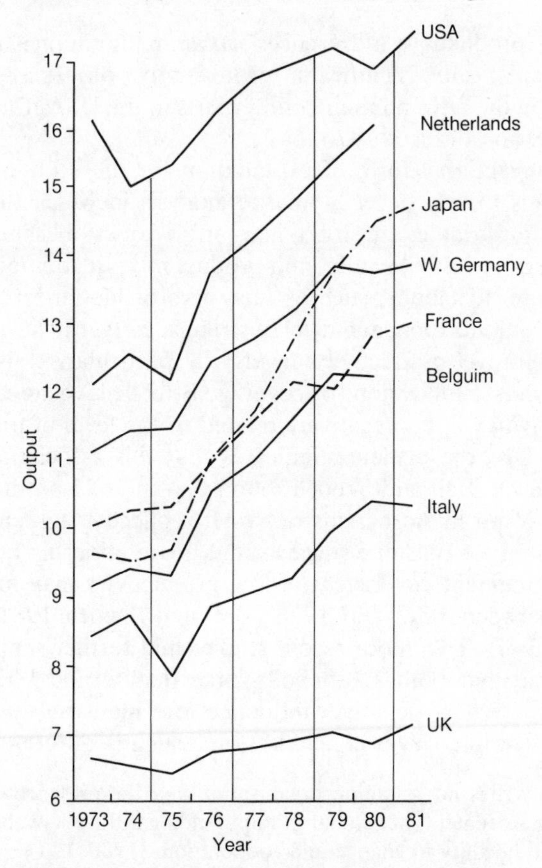

Figure 3.2. Manufacturing Output per Employed Worker-Year in Various Countries, 1973–1981
Source: Roy, *National Institute Economic Review*, 1982

tion; their comparisons with West Germany for the same year vary from 2.18 : 1 in construction to 0.82 : 1 in agriculture (1982, 5). Moreover, the generally disadvantageous comparison extends to other countries.[11] As Figure 3.2 shows, for manufacturing, UK labour productivity has been far lower than that of its major industrial competitors for some time.

Industry studies confirm the more general picture. For example, Prais finds that productivity for his ten manufacturing industries differs between the USA and UK by factors of between 2 and 5, the comparison always being in favour of the USA (1981). Pratten, analysing productivity differences *within* multinational companies, found statistically significant productivity differences between activities in the UK and their counterparts in the USA, Canada, West Germany, and France (1976, 5–8).

The conventional form of explanation for these observed differences tends to rely on econometric analysis in which the principal explanatory variables are size of market, size of establishment, capital intensity, capacity utilization, and product mix. In addition, variables which relate to labour, such as measures of labour-force 'quality', degree of unionization, and level of strike activity, can have an impact on comparative productivity levels: in particular, it is generally assumed that unionization (or coverage by collective agreement) and strike activity will be negatively related to the level of productivity.

In the UK, the evidence indicates that this association holds at industry level. Ball and Skeoch found that in 1973 productivity was generally lower in those industries where collective agreement coverage was greater; Batstone suggests that the relationship between collective agreement coverage and low productivity may have strengthened between 1973 and 1978 (Ball and Skeoch 1981; Batstone 1984, 142–4). Two more recent studies add further support. Smith *et al*. argue that both UK labour-force quality and UK industrial relations exert a negative influence on industrial performance (1982, 97), while Caves concludes, on a slightly different note,

> the difficulty lies not in union organisation and the presence of collective bargaining agreements *per se* but in longstanding attitudes of the workforce that sustain hostility to change and cooperation. (1980, 179.)

The nature of the problem is thus a little elusive: it is not the case that the UK strike rate *per se* causes productivity problems (Prais 1981, 75–9; Smith *et al*. 1982, 51), nor are productivity differences likely to emerge only from differences in degree of unionization. Rather the problem relates, as in the classic formulation by Ulman (1968, 324–80), to peculiarities of the UK industrial relations system. However, it is quite difficult to attribute independence in different degrees to variables which are so classified within a regression analysis. As Caves himself notes, 'one retardant of productivity tends to indulge or promote another, so that the whole is, as it were, less than the sum of its parts' (1980, 143). The disadvantages of the UK industrial relations system thus coexist with observed problems of managerial competence, under-capitalization, and a failure in some circumstances to achieve scale economies: labour problems may stem from this latter set, or they may contribute to it (Jones and Prais, 1978).

The relevance of this discussion to problems of technological change is indirect. One can only attribute the productivity problem to labour or trade union behaviour when other things are equal; in international productivity comparisons they manifestly are not. Nevertheless, some aspects of the British productivity problem are of direct interest. Smith *et al*. suggest that UK relative performance is poor in

(i) capital-intensive industries;
(ii) activities where the productivity potential is highest;
(iii) industries where plant size is typically large. (1982, 85, 96.)

However, as Bell (1983, 4) points out, it is one thing to identify the factors which influence productivity levels, and quite another to isolate those which influence the rate of productivity growth. One needs to show that the UK's failure to adopt best-practice techniques which might 'solve' the productivity problem is in some way related to labour resistance.

Once more, there is indirect evidence from productivity data. One way to measure the economic impact of technology is by the rate of advance of best-practice productivity (Maddison 1979, 202),[12] and to make the further assumption that 'productivity in the United States is highly correlated with the maximum level of productivity attainable using the technological knowledge and management practices now available to the industrial countries' (Caves 1980, 157). The rate of growth of productivity in the USA, or other countries close to it, depends on the technological frontier, while from this viewpoint productivity growth in countries such as the UK depends on the capacity to imitate and adapt best practice developed elsewhere[13] (Maddison 1979, 195; Lindbeck 1983). The failure of the UK to close the gap, even with the 1970s productivity slow-down in the USA, may reflect the difficulties noted by Smith *et al*. Indeed, for the period of 1963–73, Wragg and Robertson found that, within the UK, growth in *total factor* productivity was *lower* in industries experiencing more strikes per worker, and negatively related to collective agreement coverage by industry (1978, 54–7). There are examples from the steel industry, national newspapers and can-making of labour-related long-term loss of output from new facilities (Ayton 1982; Martin 1981; Prais 1981), but, once more, there are problems with this kind of data. On the one hand, it may be unrepresentative of events elsewhere; on the other, the analysis does not allow one to separate out the effect of labour resistance from other possible sources of productivity reduction.

As the next chapter will show, this approach tends to adopt a rather restrictive definition of technical change; as Abernathy (1978) has noted, improvements in productivity and the achievement of large-scale technical change may be incompatible within a given industry. Nevertheless, at the macro-economic level, it seems clear that technological change contributes to productivity growth (Denison 1967) and that the UK's failure to innovate, noted in the previous section, and the productivity problem noted here are compatible. It also seems likely that employee behaviour and attitudes have acted as a brake on both innovation and efficiency. However, econometric analysis sheds no light on the mechanisms through which this brake acts, and in order to identify such mechanisms one must pursue a disaggregated analysis which focuses on individual firms and industries.

INEFFICIENT LABOUR UTILIZATION

For many economists, productivity restrictions which originate with labour fall under the heading of 'restrictive practices'. These have been defined by Ulman to include

> Excessive job fragmentation by craft unions: overmanning: double handling: excessive tea breaks or other forms of idle time: retention of the same number of machines per operative following the introduction of new and improved machinery: limiting (or 'pegging') output per worker under either time or incentive payment: limiting output of equipment: excessive overtime. (1968, 340.)

In short, the approach includes any practice which may inhibit efficiency, and presumes that such practices originate with collective labour organization which has less interest in efficiency than have managers of industrial concerns.

Ulman's prime concern is with efficient utilization of labour rather than technological change. As the discussion above notes, more efficient utilization of existing equipment may be an alternative to investment, but in many cases technological change and efficiency improvements are related. For one thing, 'Old fashioned plant and equipment furnishes a most hospitable environment for the perpetuation of old fashioned working habits and related institutional arrangements' (Ulman 1968, 327). As a corollary, technical change provides the opportunity for rooting out inefficiencies in labour utilization: Slichter *et al*. refer to technological change as 'the most effective . . . weapon' against restrictive practices in the USA (1960, 333), while in the UK McKersie and Hunter find a weak positive relationship between the capital intensity of an industry and its coverage by productivity bargaining.[14] Such bargains were frequently addressed to the removal of practices which could be seen as hangovers from a *previous* technology, and could themselves break the circle of low wages and obstruction of change, making it doubly attractive to firms to invest in more capital-intensive processes (McKersie and Hunter 1973; Ch. 3, 4; Ulman 1968; 365–6).

Certainly, the coverage of such agreements gives some clue as to the extent of restrictive practices in the UK in the 1960s. McKersie and Hunter identify 73 productivity agreements across 12 industrial sectors between 1963 and 1966, and over 4000 across all 24 SICs between 1967 and 1970 (1973, 45,67). The most prominent industries in the first phase were chemicals, engineering, and transport, and in the second, engineering, food, and paper. These agreements may not have removed inefficiencies in the use of labour (i.e. a certain proportion may have been 'bogus') but it seems reasonable to suppose that at least the majority resulted from some concern about the pattern of labour utilization, whether through management oversight or the employees' commitment to particular practices.

Pratten and Atkinson have summarized the results of a number of different studies of inefficient labour utilization in the UK (1976, 571–6); their results are presented here as Table 3.3. Subsequently, Pratten's own work on productivity differences within multinational companies has remedied one of the defects identified in these earlier studies: namely, that they do not quantify the effects of different sources of inefficiency. His respondents felt that strikes and restrictive practices contributed about one-tenth of the productivity shortfall of UK operations on those in the USA—approximately the same fraction as that accounted for by machinery differences. As noted above, over one-fifth of companies experienced restrictive practices in the UK which they did not encounter in the USA (Pratten 1976, 61, 83–105). Other quantifiable productivity consequences of inefficient labour utilization are provided by the much-quoted

Table 3.3. Industry Studies of

Industry and report	Date of report	Does over-manning or inefficiency exist?	Sections or trades to which over-manning or inefficiency applies
Shipbuilding			
Geddes	1966	Yes	General
Alexander, *Fairfields*	1970	Yes	General
Department of Trade and Industry	1971	Yes	General
Hill Samuel	1971	Yes	
CIR Report No. 22	1971	Yes	General
Booz-Allen	1973	Yes	Outfitting trades
Chemicals			
Anglo-American Prod. Report	1953	Yes	Maintenance
Flanders, *Fawley*	1964	Yes	General
NEDO, *Manpower in the Chemical Industry*	1967	Yes	Maintenance
NEDO, *Chemical Manpower in Europe*	1973	Yes	Maintenance
Printing			
Anglo-American Prod. Reports			
(1) Letterpress	1951	Yes	General
(2) Lithograph	1951	Yes	General
The Economist Intelligence Unit	1966	Yes	General (particularly machine rooms and process dept.)
NEDO, *Printing in a Competitive World*	1970	Yes	Bridging and finishing
Motor vehicles			
Maxcy and Silberston	1959		
Clack	1967		
Beynon	1973	Yes	
Ryder Report[c]	1975	Yes	General
Mechanical engineering (incl. machine tools)			
Way Report	1970	Yes	
NEDO Industrial Review to 1977	1973	Yes	
Textiles			
The Platt Mission	1944	Yes	
Anglo-American Prod. Reports, *Cotton, Spinning, Yarn Doubling, and Cotton Weaving*	1950	Yes	
The British Cotton Industry	1952	Yes	
Caroline Miles, *Lancashire Textiles*	1968	Yes	
The Textile Council, *Cotton and Allied Textiles*	1969	Yes	

Notes:

• Indicates that the cause was mentioned in the report.

•• Indicates that this cause of differences in performance was emphasized.

Inefficient Labour Utilization

Causes of overmanning or inefficiency					Other causes of low productivity			
Labour restrictive practices[a]	Union structure	Strikes	Unsatisfactory management–union negotiating machinery	Management failures	Scale difference size of plants production runs etc[b]	Old vintages of capital equipment	Other differences in capital equipment	Shift working restricted
•	•	•		••				
•			•	••				
•			••			•		•
	•			•				
•	•	••	••					
			••	••	•	••		
•				••	•			
•			•	••				
•				••	••			
•	•			••				
•			••	••	•			
•				••				
•	••		•	••				
	•			•	•			
				•	•			
		•						
•		•	•	•				
•	•	•	•	•	•	•	•	
•				•	•		•	•
				•	•		•	
				•	•	•		
				•	•	•	•	•
				•	•	•	•	
					•	•		•
				•	•	•		•

[a] Restrictive practices include demarcation rules, especially between crafts and non-craft workers, and union rules involving the use of mates.
[b] This heading includes differences in vertical integration.
[c] Also the CPRS Report on the *Future of the British Car Industry*, HMSO, 1975.

Source: Pratten and Atkinson 1976.

CPRS study of the UK car industry: on identical equipment, continental plants produced over twice as much output per man (CPRS 1975, 83–85). Other work has illustrated that restrictive practices have developed in engineering (Brown 1973), dockwork (Mellish 1973), and metal-box manufacture (Willman 1982a), or that industries identified by Pratten and Atkinson have continued to be so afflicted for some time (for example, Martin (1981) for newspapers and Willman and Winch (1985) for the car industry).

The point needs to be made, of course, that the existence of restrictive practices does not *of necessity* signal trade union or employee resistance to technological change or efficiency improvements. It is the essence of the classical analysis of feather-bedding 'that it results from the attempt by trade unions to "carry forward" a set of practices appropriate for one technology to another where it is alien' (Weinstein 1964, 147), but a *non sequitur* then to argue that overmanning follows solely from trade union attitudes or behaviour. The data from Pratten and Atkinson presented in Table 3.3 reveal different attributions of blame for inefficiency. Both Prais (1981, 137), discussing the food industry and Gomulka (1979, 176), quoting evidence from engineering, illustrate inefficiencies in labour and capital utilization which do not stem from trade union work rules. Pratten (1976) does not quote such examples, but this methodology is unlikely to have unveiled them.[15] Moreover, attribution of cause notwithstanding, it is by now axiomatic for students of industrial relations that rules of custom and practice (which may be the source of such inefficiencies) are premised on some form of managerial action or oversight (Brown 1973; Batstone *et al.* 1977; Terry 1977). If restrictive practices are widespread in the UK, their removal depends as much on the modification of management behaviour as on the eradication of trade union resistance.

In fact, managerial behaviour does appear to be extremely important. Firstly, there are the problems of managerial suspicion and resistance to change: this may be a particular problem where managers lack technical expertise (Hutton and Lawrence 1981). Of more central concern here is the argument advanced by Prais (1981, 260–2) that managers become too involved in day-to-day labour relations issues to consider efficiency or innovation. This suggests a rather unfortunate cycle: managers cannot consider change because they are involved in continuous negotiations the outcome of which is a set of rules which further inhibit efficiency.

This difficulty may be compounded by the way in which managers handle labour-displacing change. Comparing the UK and West Germany, Jacobs *et al*. (1978, 115–18) suggest that German managers take much more account of employment stability and devise more thorough strategies for its management than their UK counterparts. As a result, British workers tend to have greater anxieties about employment than their German counterparts. This work supports the implication of the discussion in Chapter 2: namely, that collective bargaining in the UK seems comparatively inefficient in the negotiation of change.

More recently, there appears to have been an acceleration in the reform of working practices in UK manufacturing; occurring in a period of high unemployment, it does not appear to have occasioned much labour resistance (Batstone 1984, 242; Edwards 1984). The capacity of trade unions to resist change appears to have reduced, and managers appear to have addressed problems of productivity and efficiency more rigorously. However, if there has been no reform of management practice upon which to base such changes, then they may not last. Bell (1983, 34) suggests that the UK's weak international competitiveness and the inability to offer job security may be related and reinforcing, in that trade union attempts to create security in the short term through productivity-inhibiting devices may further weaken the competitive position of their employers.

A focus on managerial strategy and the place of technological innovation in the search for high productivity or improvements in efficiency is thus central to the issues being discussed here. It is plausible to suggest that firms operating in different product- and labour-markets experience competitive pressures of differing types and severity, and that the responses to such pressure will consist of a varying mix of cost-cutting process innovation, product innovation, or organization innovation for improved efficiency. The next chapter will seek to outline the major features of these different circumstances.

CONCLUSION

The evidence presented in this chapter indicates that the UK does suffer from some sort of 'innovation problem'. In terms of both R&D expenditure and US patents, innovative activity in the UK does not match that of a number of our international competitors. There is some evidence of opposition to change by trade unions, but this is insufficiently extensive to justify the view that UK trade union behaviour is a major obstacle to technological change. Innovation is not closely associated with strike activity, but the UK's low productivity does appear to be related to inefficiencies in the utilization of labour associated with the retention of outmoded working practices: these may not wholly be the responsibility of trade unions. In fact, the failure to achieve rapid rates of innovation appears to be associated with the more general failure to achieve efficiency. Many restrictive practices, particularly those associated with manning levels, survive because of failure to achieve efficient labour utilization after previous generations of technological change. This failure may be real, in that goals of efficiency have not been reached, or apparent, in that the objective of innovation might not have been simply efficiency.

The available evidence does enable us to define the 'problem' more precisely. Although there has been much less academic investigation of productivity in the service sector or of white-collar union activity, it does seem that resistance of all forms is more likely to be associated with manual workers and to occur in manufacturing and transport. The incidence of strike activity on the rather imperfect measures used here is even narrower, encouraging a focus on docks, national newspapers, and car production.

To go further in the analysis of resistance to technical change one needs to look at the independent variable, so to speak: the nature of change in different industries. The distinction between process and product innovations is important here. If the purpose of the former may be generally said to be movement of cost curves downwards, the purpose of the latter is often to shift the demand curve outward. The productivity consequences may differ markedly in the short term. The extent of process and product innovation is thus important for trade union behaviour. As Chapter 2 illustrated, such shifts in demand curves are generally supported by trade unions and, as I shall show below, product innovations are unlikely to be contested. But it does not thereby follow that process change will always occasion resistance. In order to discriminate between those cases where it does and those where it does not, one must look more closely at the nature of innovation within particular firms and industries.

NOTES

1. The table presenting these results also appears in Freeman (1979).
2. The corresponding figures for West Germany and Japan in 1979 are 23.9 and 27.7 per cent respectively.

3. In each of the sub-periods 1870–1913, 1913–50, 1950–70 and 1970–7, UK rates of growth were below the average of the USA, Japan, Italy, Germany, and France. Their figures are drawn from Maddison (1979) and Hill and Utterback (1979).
4. This result has not been more generally reproduced: see Prais (1980, 195–6).
5. This causes them to doubt the reliability of their capital intensity data.
6. This does not, of course, imply that labour costs per unit of output are low in the UK: see Roy (1982), Ray (1984a).
7. As Durcan *et al*. note, the classification has itself changed: 'the department began in 1946 listing eight principal causes, reduced this to seven in 1953, then extended it to nine in 1959' (1983, 17).
8. This evidence arises from a comparison of the SPRU figures on process innovations in manufacturing (Figure 2.1) with the figures on major strikes by cause, presented by Durcan *et al*. The SPRU data reveal that the period 1947–60 was one of rapid *increase* in rates of process innovation and the period 1960–75 one of overall *decrease*. Durcan *et al.*'s figures show the following percentages of strikes 'caused' by redundancy and demarcation: 1946–52, 12.2 per cent; 1953–9, 15.9 per cent; 1960–8, 8.7 per cent; 1967–73, 6.0 per cent.
9. For example, over the period 1978–82, the numbers and proportions were:

Year	No. of prominent stoppages	% of w.d.l.
1978	221	81.5
1979	254	95.0
1980	100	93.0
1981	105	78.3
1982	101	82.8

10. Rs = 0.1935; the relationship is not significant.
11. In a matched sample of Swedish and UK companies, Pratten also found labour productivity differences of about 50 per cent (Pratten 1976, 131 f.).
12. Another is to estimate residuals on production functions and to attribute them to technical advance (see Solow 1957, Jorgensen and Griliches 1967).
13. This is essentially the view of Gomulka (1979, 175) who argues that the long-run rate of technical change is approximately equal to the growth rate of labour productivity.
14. They assume a relationship between capital intensity and historical experience of technical change (McKersie and Hunter 1973, 358).

Ulman suggests that this sort of relationship reflects the requirement to minimize down-time on expensive equipment by negotiating maintenance flexibility (1968, 359).

15. His respondents were primarily managerial employees of multinational companies.

4

The Pattern of Industrial Innovation

INTRODUCTION

THE previous two chapters have tended to treat innovation as a relatively homogeneous concept. However, as the general discussion of Chapter 1 indicated, a number of distinctions between different types of innovation are not only helpful but necessary. Innovations may affect production processes, or products, or both; in addition, they may be radical or incremental. Other distinctions are necessary to analyse innovation within particular firms. Companies may use the latest technology in process and product innovations, or they may simply 'borrow' existing technology from competitors: by implication, then, change may originate within a particular company or be imported from without. In short, innovation must be considered in terms of the overall strategy pursued by companies in different markets: it may form part of the strategy for growth or product-market success, or it may act as an external constraint, altering the direction a particular organization seeks to take. Important questions thus concern 'the relationships of technological progress to changes in other factors: productivity, innovation, production organization, work force skills, advances in production equipment and new material sources' (Abernathy 1978, 4). These relationships may differ between industries or firms; they may also change over time. However, the essence of the approach I shall describe is that it seeks to locate the process of innovation within the overall strategies pursued by particular companies.

The purpose of this chapter will be to present an integrative model of innovation, developed in a number of papers by Abernathy, Utterback, and others, and to spell out in some detail its implications both for company labour relations strategies and for trade union responses to change. I shall seek to show that a particular subset of process innovations in mature industries is likely to be of particular concern to trade unions, given their policy goals outlined above, and that these process innovations are thus likely to be among those which occasion some form of resistance.

THE ABERNATHY–UTTERBACK MODEL

In the approach presented here, innovation refers to the first use or application of a particular technology in a given industry (Utterback 1974, 621). It is thus distinct from invention or technical prototype, and indeed the lead time between invention and innovation may be substantial (Freeman *et al*. 1982, 51–6). The integrative properties of this essentially descriptive approach stem from its capacity to delineate systematic relationships over time in the development of an industry between process and product innovation and between radical and incremental change; furthermore, it discovers coherent patterns of change over time both in the stimuli for innovations and the barriers to them (Utterback and Abernathy 1975, 639). The model has been developed in a number of publications focusing on the relationship between competitive strategy, characteristics of the production process, and innovation (Abernathy and Wayne 1974; Utterback 1974; Abernathy and Townsend 1975; Utterback and Abernathy 1975; Abernathy and Utterback 1982; Abernathy 1978; Utterback 1979). The basic features are illustrated in Figure 4.1 and Table 4.1; a number of general points need to be made at the outset.

The first is that the approach largely sidesteps debates about general 'demand-pull' and 'technology-push' effects upon *rates* of innovation by arguing that the stimuli towards innovation alter over time and that process and product innovation feed upon one another to set up the conditions for evaluating development through different stages.[1] The second is that the unit of analysis is not necessarily the firm or industry but the 'productive unit', i.e. 'a product line and its associated production process' (Abernathy and Utterback 1982; 102).[2] The third is that, because a range of feasible strategic motives for innovation are considered, the assumption that innovation and productivity improvement are highly correlated does not always hold. In fact, it is central to the approach that for 'mature' industries there will be a clear trade-off between radical technical change and productivity improvement: this has been empirically demonstrated in some detail in those studies focusing on the motor car industry (Abernathy and Wayne 1974; Abernathy 1978).

A succinct description of the relationship outlined in Figure 4.1 and Table 4.1 runs as follows.[3] An industry begins through the origination of a major product innovation, which is more likely to be demand-stimulated than a consequence of technological change *per*

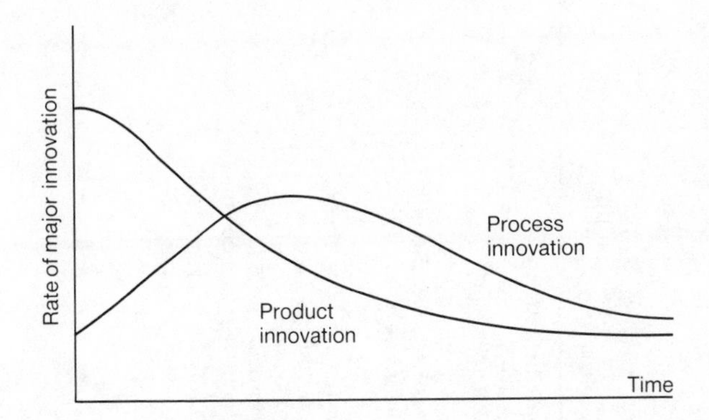

Figure 4.1. A Model of Process and Product Innovation

se. Product innovations in the early stages are rapid, with a high failure rate and substantial market redefinition. Processes, too, tend to change rapidly; they remain fluid and responsive to change yet relatively inefficient. Innovation tends to come from views about product users' needs, and firm size tends to be small. Processes are characterized as 'unco-ordinated' and inefficient; the basis of competition is maximizing product performance rather than product price.

In the second phase, advanced technology becomes much more important for both product and process. As product designs stabilize and a 'dominant' design emerges, the competitive strategy shifts to sales maximization:

> As obvious improvements are introduced, it becomes increasingly difficult to better past performance, users develop loyalties and preferences and the practicalities of marketing, distribution, maintenance, advertising, etc. demand greater standardization. (Utterback and Abernathy 1975, 644.)

Processes become 'segmental' as demand for increased output involves the design of more rigid and mechanistic production systems concerned with efficiency: the production system involves specialized tasks, tighter controls, and more automation. Advanced technology also becomes the basis for product differentiation.

In the third phase, product and process designs become more closely linked. High levels of product standardization induce a shift of competitive emphasis towards product price: margins are reduced, there are tendencies towards oligopoly, and economies of scale become important. Processes become 'systemic' and rigid: the

Table 4.1. A Model of Process and Product Innovation (see Figure 4.1)

	Fluid pattern	Transitional pattern	Specific pattern
Competitive emphasis on	Functional product performance	Product variation	Cost reduction
Innovation stimulated by	Information on users' needs and users' technical inputs	Opportunities created by expanding internal technical capability	Pressure to reduce cost and improve quality
Predominant type of innovation	Frequent major changes in products	Major process changes required by rising volume	Incremental for product and process, with cumulative improvement in productivity and quality
Product line	Diverse, often including custom designs	Includes at least one product design stable enough to have significant production volume	Mostly undifferentiated standard products

Production processes	Flexible and inefficient; major changes easily accommodated	Becoming more rigid, with changes occuring in major steps	Efficient, capital-intensive, and rigid; cost of change is high
Equipment	General-purpose, requiring highly skilled labour	Some subprocesses automated, creating 'islands of automation'	Special-purpose, mostly automatic with labour tasks mainly monitoring and control
Materials	Inputs are limited to generally available materials	Specialized materials may be demanded from some suppliers	Specialized materials will be demanded; if not available, vertical integration will be extensive
Plant	Small-scale, located near user or source of technology	General-purpose with specialized sections	Large-scale, highly specific to particular products
Organizational control is	Informal and entrepreneurial	Through liaison relationships, project and task groups	Through emphasis on structure, goals, and rules

Source: Abernathy and Utterback 1982.

emphasis is upon incremental non-disruptive small-scale change since radical process change involves massive market and productivity loss. However, market needs are well articulated and lend themselves to complex technological solutions; and

> Process redesign may be spurred either by the development of new technology or by a sudden or cumulative shift in the requirements of the market. If changes are resisted as process technology and the market continue to evolve, then the stage is set for either economic decay or a revolutionary as opposed to evolutionary change. (Utterback and Abernathy 1975, 642.)

This broad descriptive model is useful in a number of respects for the present purpose. It disaggregates the notion of innovation in such a way as to allow its differential impact on industrial relations to be understood. It serves to emphasize the variety of motives for innovation: during the second phase, technology is the pacing element in strategy; in the third, it can be substituted by other cost-minimizing devices. The importance of inter-industry and inter-firm diffusion mechanisms is also outlined, again in a systematic way. However, perhaps the most important feature of the approach is that it contains a rudimentary treatment of the relationship between innovation and organizational structure which on the face of it is compatible with more rigorous work in organizational theory (Burns and Stalker 1961; Lorsch 1965).

Some elaboration is necessary. Abernathy and Utterback's major concern is with company innovation strategy rather than labour policy or trade union response; their model is synthetic rather than innovatory in itself and, as Table 4.1 shows, the suggested organizational responses are presented in a sketchy and routine fashion. Moreover, it is based almost exclusively on US data and its more general applicability may be in doubt. It is thus necessary, firstly, to demonstrate the relevance to the UK and, secondly, to spell out the implications of this approach for the concerns discussed in the previous three chapters.

INNOVATION IN THE UK

Although the Abernathy–Utterback model is based on US data, attempts have been made to classify UK industries according to type of innovative activity, and its relationship to competitive strategy. Freeman, for example, distinguishes between:

(i) capital goods and chemical industries which spend a great deal on R&D, directed mainly towards development of superior new products;
(ii) basic materials industries (such as rubber and metal products) which spend proportionately rather less, mainly on factor-saving and cost-reducing efforts;
(iii) consumer goods industries which spend very little, but adopt cost-cutting innovations from the high R&D capital goods sectors. (1979, 59–60.)

In a similar vein, de Bresson and Townsend (1978) have sought to develop a classification of UK industries according to competitive strategy and innovation. Following Simmonds (1973) they distinguish 'performance-maximizing', 'sales-maximizing', and 'cost-minimizing' industries in a clear parallel to the Abernathy–Utterback approach; in addition, they develop an industrial classification based upon the International SIC. Their concern is primarily with inter-industry flows of technology, and they are able to show the predominance of the performance-maximizing sector as a source of innovation: it provides 56 per cent of all innovations to the two other sectors, and 95 per cent of its own. By contrast, cost-minimizing industries are receivers of innovation: they use 43 per cent of all innovations, but produce only two-thirds of their own (1978, 50–3).

This approach can be extended. Table 4.2 presents an adaptation of the original classification to the UK SIC.[4] Some considerable inaccuracies are involved in this type of approach: not only is the SIC not sufficiently disaggregated to permit finer distinctions between industries but, it will be recalled, the unit of analysis in this type of approach is neither the firm nor the industry but the 'productive unit', which might cross the boundaries of both. In addition, the approach is not exhaustive and focuses primarily upon manufacturing and transportation; other types of industry included are resource-based (i.e. extractive), distribution, finance, and capital goods, in all of which sales-maximization or cost-minimization considerations are important. Nevertheless, the classification is of some interest in distinguishing broad areas in which different types of innovation might be expected. The usefulness of the classification is shown in Table 4.3. The table uses the SPRU data base on innovations in the UK in the post-war period classified according to Table 4.2 (Townsend *et al.* 1981). As the Abernathy–Utterback approach would imply, product

Table 4.2. Innovative Sectors in the UK

Sector type	Industry	1968 SIC
Cost-minimizing	Textiles; Wood products; Paper and paper products; Bricks, pottery, china, earthenware; Glass; Other non-metallic minerals; Iron and steel; Other basic metals; Fabricated metal; Transport equipment; Electricity, gas, and steam; Water works; All transport; Communication	SIC Order XIII, XVIII, XVI, VI, XII, XXI MLH 471, 474, 475, 479, 370, 384, 385 SIC XXII
Sales maximizing	Automobiles;[a] Detergents; Cosmetics; Food, drink, tobacco; Clothing; Footwear; Furniture; Other manufacturing; Retailing; Restaurants; Hotels	SIC III, XIV, XV MLH 273, 275 MLH 381, 382, 493 MLH 472, 473, 494, 495, 499 MLH 820, 821, 884, 885, 886, 887
Performance-maximizing	Chemicals (industrial and other), petroleum; Rubber, plastics; Non-electrical machinery; Electrical machinery; Scientific instruments; Aerospace[a]	SIC IV, VII, VIII, IX MLH 380, 383, 491, 492, 496 MLH 271, 272, 274, 276–9

[a] The UK SIC allows this to be separated from other transportation equipment which the ISIC does not.

Source: De Bresson and Townsend 1978.

Table 4.3. Process and Product Innovations by Industrial Category

Sector type	Average	
	% Product	% Process
Performance-maximising (*n* = 19)	76.4	12.9
Sales-maximizing (*n* = 4)	57.2	39.9
Cost-minimizing (*n* = 7)	62.0	27.1

Source: SPRU data, from Townsend *et al*. 1981.

innovation is of the greatest importance in performance-maximizing sectors, while process innovations numerically predominate in sales-maximizing ones. The classification does appear to distinguish different innovative patterns in the UK.

These are, of course, very general categories which themselves will almost certainly contain substantial innovative variance. However, a number of important differences of relevance to the concerns here do exist between sectors: some of these are summarized in Table 4.4 which relies on data in a sample of manufacturing industries originally presented in Wragg and Robertson (1978). Table 4.4 shows mean compound rates of growth of output, employment, unit wages, and two measures of output per head, reclassified according to sectorial competitive strategy. The performance-maximizing sector shows employment growth, rapid rise of output per head, and low rate of growth of unit wage costs. To judge by the performance of the other two sectors, progression along the Abernathy–Utterback developmental curve in post-war UK has involved high rates of employment contraction and higher-than-average increase in unit wage costs in return for small gains in output per head. On all five measures the performance-maximizing sector differs significantly from the other two, which are themselves relatively similar.

Wragg and Robertson's data go only as far as 1973, but later data from the 1979 Census of production confirm that unit wage costs are markedly lower in performance-maximizing industries than elsewhere. In a sample of 164 industrial categories classified by competitive sector, wages per head varied by less than 25 per cent; however, average value added per head in performance-maximizing industries

Table 4.4. Rates of Growth by Sectors in Output, Employment, Wages, and Output per head

Variable	Compound rate of growth (% p.a.)[a]			
	$\bar{x}$ Performance-maximizing sector (n = 23)	$\bar{x}$ Sales-maximizing sector (n = 34)	$\bar{x}$ Cost-minimizing sector (n = 23)	$\bar{x}$ Overall (n = 80)
Gross output	4.28	1.85[b]	1.55[b]	2.46
Total employment	0.07	−1.49[b]	−1.48[b]	−1.04
Output per head	4.43	3.18[c]	3.22[c]	3.55
Output per operative	4.88	3.77[c]	3.48[b]	4.00
Unit wage	2.69	3.60[c]	4.08[c]	3.48

Notes:

Wragg and Robertson include two 'resource-based' industries which are excluded here.

[a] $\bar{x}$ = average.

[b] Significantly different from performance-maximizing sector beyond 0.01.

[c] Significantly different from performance-maximizing sector beyond 0.05.

Source: Wragg and Robertson 1978.

was 37.7 per cent greater than the average for sales-maximizing industries and 64.9 per cent greater than the average for cost-minimizing ones. These figures do not, of course, control for differences in the level of capital investment per head.

These findings are of some importance given the relevance of changes in the economic climate for trade union policy illustrated in Chapter 2. *Ceteris paribus*, the scope for trade unions to accept change and for employers to concede wage rises in return would seem to be greatest in the performance-maximizing sector, which is in turn the sector where the greatest volume of product innovation will occur. By contrast, in the sales-maximizing and cost-minimizing areas of manufacturing, employment contraction has gone along with a more rapid rise in unit wage costs than in output per head, pointing to the possibility of greater difficulties in the negotiation of change: these are the sectors where process innovation is most likely.

The Abernathy–Utterback approach is not without its critics. In particular, the description of shifts between product and process innovation over time does not characterize developments in many process industries. Chemical and petrochemical industries appear to have begun with radical process innovations, to have engaged in product diversification, and to have reverted once more to process innovations (de Bresson and Townsend 1981, 435). Even within engineering, small specialized supplier firms with high levels of product innovation appear to persist over time. Pavitt (1984), seeking a classification of innovative patterns on the basis of SPRU data, thus divides firms between 'science based' and 'supplier dominated' categories, which broadly correspond to performance-maximizing and cost-minimizing sectors, but divides his third category between 'scale intensive' and 'specialized supplier' firms. The former, exemplified by large firms in steel, vehicle production, and consumer durables, are process innovators; the latter, in machinery and instrumentation, are product innovators.

INNOVATIONS AND RESISTANCE

The classification may also serve to distinguish industrial sectors in terms of their experience of conflict over change. Table 4.5 presents data on employment change and unionization for the three sectors, covering in this case both manufacturing and services. It shows that the performance-maximizing sector has experienced rapid employment growth from a lower base, and that it has relatively high union density and a relatively high incidence of industrial conflict; within this sector, engineering industries account for a much greater proportion of industrial conflict and show higher union densities, as well as much higher growth of employment and unit wages, but lower output growth than the chemcial industries.[5] When the sales-maximizing sector is extended outside manufacturing, it shows moderate employment growth and very low levels of unionization and conflict. Cost-minimizing industries are highly unionized, conflict-prone, and contracting.

This pattern is of considerable interest. Both the major producers and the major receivers of innovations according to the SPRU survey have relatively high levels of unionization and conflict. The sector which, in theory, has the highest rate of process innovation is relatively low on both, although the inclusion of retailing is important here.[6] The cost-minimizing sector is particularly strike-prone. In Durcan *et al.*'s analysis of post-war strike activity, six of the top seven strike-prone industries outside coal mining—port and inland water transport, shipbuilding, road passenger transport, other transport and

Table 4.5. Sectoral Characteristics

Characteristic	Performance-maximizing sector	Sales-maximizing sector	Cost-minimizing sector
Employment[a] 1979 (000's)	2542.5	4067.3	4306
Employment change[b] 1948–79 (%)	+48.4	+13.4	−17.8
Working days lost[c] per 1000 workers 1947–73	266.9	63.8	471.5
Union density[d] 1979 (%)	74.3	58.8	81.2

Sources:

[a] For production industries, *Report on the Census of Production*, London: HMSO, 1979, Table 2, 40–58. For non-production industries, DE *Gazette*, May 1980, Quarterly estimates by MLH.

[b] For production industries, from 1970, DE *Gazette*; Prior to that *Historical Record of the Census of Production, 1907 to 1970*, London: HMSO, 1978, Table 6, 202–42. For non-production industries, *Historical Abstract of British Labour Statistics*, London: HMSO, 1971, Table 125. The former is based on the 1968 SIC, while the latter is not; there is thus some inaccuracy in the comparison for non-manufacturing elements of labour in the sales-maximizing and cost-minimizing sectors.

[c] Durcan *et al*. 1983, 176. Their coding (p. 436) suggests an attempt to maintain continuity with changes to the SIC but gives no details of the basis of their classification. Industries have been allocated to sectors on a judgemental basis.

[d] Price and Bain 1983. The basis of industrial classification of Bain and Price (1980) again involves some mismatch with that used here. In particular, 'Metals and Engineering', 'Chemicals', 'Timber and Furniture', and 'Distribution' straddle the sectoral classification. The solution adopted was to transfer 1979 employment in MLH 381, 382 from the 'metals and Engineering' (performance-maximizing) sector to the sales-maximizing sector on the assumption of 100 per cent membership; to classify all of 'Chemical' under performance-maximizing; to split 'Timber and Furniture' evenly between the sales-maximizing and cost-minimizing sectors; and to omit 'Distribution'. (Note, however, that, if Bain and Price's 'Distribution' sector is *all* included under sales-maximizing, the figure of 58.8 per cent shown drops to 34.4 per cent.)

communications, printing and publishing, and iron and steel—experience cost-minimizing competitive conditions. Moreover, as we have seen, two of the three industries experiencing the largest conflicts directly over technological change are cost-minimizing. The exception in both cases is motor vehicles.

The classification serves to emphasize the importance of product-market competition for innovation and conflict. However, generalizations across sectors should not be allowed to disguise important differences *within* sectors. For example, retailing experiences lower union densities and strike losses than other parts of the sales-maximizing sector, while the chemicals industries tend to be less strike-prone than engineering.[7] These differences need to be borne in mind in applying the approach to the study of labour relations, but three broad issues to arise from the discussion so far.

The first concerns the brief discussion of innovation motives in Chapter 1. The Abernathy–Utterback model implies a patterning of such motives from new product launches in the earlier stage to cost-cutting process changes in the third. In the second stage, process and product change and the application of high technology are both at a maximum. According to the model, adoption of new technology to differentiate products and maximize process output is important. The costs of not innovating seem to be high, and the benefits of successfully doing so might be evident in output expansion, economic success, and, perhaps, employment growth. In the third phase, technological change competes with other means of cost reduction, for example, more efficient use of existing facilities. A concern with new technology *per se* is more likely to be characteristic of performance-maximizing and sales-maximizing industries. The exception to this is the case where so-called 'de-maturity' occurs. I shall discuss this in Chapter 8.

The second issue concerns the size of plant. According to Utterback and Abernathy, firm size is greater in sales-maximizing and cost-minimizing industries than in performance-maximizing ones (1975, 654): the implication of their analysis is that plant size will also increase,[8] as will production-unit size, with attendant organizational inflexibilities. Given the discussions of Chapter 3 on the relationship between plant size and strike propensity, it may be that the UK is disproportionately disadvantaged in achieving successful change in just those industries where process innovation is likely to be high because of the difficulty of managing large organizations in the UK (for example, see Prais 1981, 262). *Ceteris paribus*, economies with idiosyncratic barriers to innovation in sales-maximizing areas would clearly experience a depressed rate of process innovation overall, developments in performance-maximizing sectors notwithstanding.[9]

The third issue concerns the importance of incremental innovation in cost-minimizing areas and its relationship to the existence of inefficiencies in UK industry. According to the Abernathy–Utterback approach, incremental innovations are stimulated by pressures to reduce costs and improve quality as products become less differentiated; they are particularly important in efficient and capital-intensive production processes involving large-scale plants (Utterback 1979, 53). In the work of Abernathy and Utterback, a range of literature is assembled to illustrate the greater overall importance of incremental over radical change in the reduction of cost or improvement in performance in industries such as automobiles, rayon, rocket engines, computer memories, light bulbs, railways, and petroleum refining (Utterback 1979, 53–6; Abernathy and Utterback 1982, 99–100). However, as Ulman has demonstrated, these types of change are just those which, in the UK, are likely to be frustrated (in the sense that there is a high likelihood that cost reductions will not emerge or that performances improvements will be costly) where restrictive practices spread from old to new equipment as a consequence of piece-meal 'defensive' investment additions (i.e. the grafting of marginal equipment improvements onto old capital stock) (Ulman 1968, 327). Union activity, incremental change, and the search for efficiency may be incompatible in the UK. These issues can be explained further by exploring those aspects of the approach which deal with organizational change.

INDUSTRIAL RELATIONS IMPLICATIONS

The Abernathy–Utterback approach locates innovation in the context of the broader corporate competitive strategies which in turn influence approaches to labour relations. Moreover, the approach to organizational and labour relations issues which receives rudimentary expression in it relies on existing literature in the area. The increased division of labour and emphasis on discipline implied, particularly by Abernathy (1978), rely on Bright (1958), while discussions of organizational structure rely on Woodward (1965).

This in itself is a source of difficulty. A reliance on Bright tends to focus on what Littler (1983, 136–7) has termed the 'engineering' model of development wherein successive automation of machining, transformation, and control tasks yields the sequence of change given in Figure 4.1 and Table 4.1. Once more, differences between engineering and chemical industries are important; moreover, the relevance to non-manufacturing areas has yet to be demonstrated empirically. In addition, a reliance on Woodward leads to the implication that there is a determinative relationship between the competitive stage of development and the organizational structure of a corporation: subsequent work (e.g. Child 1972) would encourage more emphasis on the role of strategic choice. Nevertheless, the effort to integrate labour relations issues into a model of strategy which emphasizes innovation does seem worth while in the context of an attempt to explain the relationship between trade unions and innovation, particularly since competitive failure, high costs of change, and 'bad' industrial relations appear to go together.

A recent advance in this area is the work of Kochan *et al*. They suggest that product-market change may provide a stimulus for disinvestment, if conditions become too unfavourable, for further investment in new technology or new plants, or for a shift in market position in terms of volume or quality. All of these changes are likely to have industrial relations consequences. The crucial issue is the extent to which labour costs are involved in competition, and this in turn is related to the product life cycle.

> The evolution from a growth market to a mature one, for example, typically forces firms to be more competitive with respect to prices. This leads them to shift their emphasis in industrial relations away from maintaining labour peace in order to maximize production to one of controlling labour costs, streamlining work rules and increasing productivity. (1984, 24.)

Others, by contrast, will seek to organize the product-market in order to take labour costs out of competition; product-market growth and change invariably jeopardize this.

Kochan *et al*. do not specifically consider production processes and their industrial relations impact, nor do they pay much attention to technical change. With the discussion above in mind, let us consider the following typifications, which relate to the different process/product combinations outlined so far.

(i) *Performance maximization*. Where new product development is the essence of competitive success, and this competition tends to be on performance rather than price, cost considerations are less salient. Moreover, where innovation tends to emerge from those most familiar with the process and product, rather than being imported from 'high-tech' areas, the skills and acquired idiosyncratic knowledge of those involved in production are important. In addition, the more successful firms at this stage expand rather rapidly and the less successful disappear. In this environment, employers are likely to value employee insight highly and to be less concerned about labour costs. Employees are likely to regard their commitment to the firm as essentially open-ended and to seek to bargain sequentially over new product development. For both sides, open-endedness is favoured because of the high level of product-market uncertainty. Many new products fail, and product life cycles tend to be short, giving ample opportunity for bargaining over the terms of co-operation with new product innovations.

(ii) *Sales maximization*. In the second stage, economies of scale become important and, as markets expand, the division of labour advances. As, under the Abernathy–Utterback model, advanced technology becomes important to process and product, the premium on idiosyncratic knowledge declines, and advances in the process of production depend upon stimuli from external 'scientific' sources. In this stage, requirements for continuity of production and co-operation with recurrent change encourage employers to attempt both to introduce predictability into the employment relationship and to secure employee acceptance of process changes. Employees may see employment expanding but task content and autonomy will change. Within individual firms, short-term product-market hiccups and consequent levelling of employment growth or lay-offs similarly require assurances about job content, work allocation, and job security.

(iii) *Cost minimization*. In the third stage, extensions to product range or expansion of sales are less likely, and consequentially most change involves some threat to employment levels or job security—particularly for direct workers. Labour productivity becomes a focus of attention and pressures for maximum capacity utilization increase. Employers seek to reduce labour costs to the lowest level consistent with continuity of production and may seek to secure the co-operation of a 'core' labour-force while offering less security and lower wages to temporary or seasonal labour. Employees press for extensions to job-security provisions and the right to negotiate over and control the implementation of incremental change. In the medium term, security is constrained by the ability to compete on cost; in the short term, it may be secured by the ability to threaten continuity of production. Overall, the acceptance of technological

Table 4.6. Employment and Output in Three Sectors

Sector type	Percentage with employment increase	Percentage with output increase
Performance-maximizing	65	100
Sales-maximizing	24	82
Cost-minimizing	38	67

Source: Wragg and Robertson 1978.

change may be encouraged by employer threats of disinvestment or closure.

These three typifications illustrate the prospect of systematic differences in the 'effort bargain'—by which I mean the relationship between employee effort input and the reward from the employer (Behrend 1957)—in that the requirements of both sides may change systematically. At the most general level, this relates to employment and the overall effort level in a given industry. Chapter 2 suggested that employment–output relationships were important influences on trade union responses to change. Table 4.6, once more using Wragg and Robertson's data, records that there were systematic differences between sectors in these terms. In the UK, those sectors which experienced the highest levels of process innovation are also likely to experience employment contraction; product innovation, by contrast, is associated with employment expansion.

One difficulty here is the extent to which one can employ the idea of a life cycle—implying necessary movement through the stages of the Abernathy–Utterback model—to characterize changes in effort bargains. Several authors have attempted to discuss similar sets of changes to organization or industrial relations. Donaldson (1985), for example, suggests a move towards divisional organization and from organic to mechanistic forms as products mature. More systematically, Hax (1985) relates human resource strategy, including labour relations, to aspects of the business life cycle, while Skinner (1974) has argued directly for the focusing of manufacturing strategy, including work organization and industrial relations, around the evolution of product strategy.

All such approaches are jeopardized by the empirical elusiveness of life-cycle tendencies in certain industries. However, a further problem concerns the relationship between product-market strategy and labour relations practice. Many large firms operate in different types of product-market, but with one overall strategy for remuneration, job security, and industrial relations, and thus do not build labour relations policies directly around product-market change. Moreover, although there has not been a great deal of research in the area, it does not appear to be the case that firms in similar product-market situations are identical in labour relations policies. As Rubery *et al.* note, product or process changes afford bargaining opportunities to employer and employees, and short-term gains for both sides may be exploited or forgone. The outcome of such bargaining will affect the extent to which employees are insulated from product-market fluctuations (1984, 105–17).

The institutional arrangements for regulating effort bargains are thus of crucial importance in times of process or product change. Firms may vary both in their *capacity* and in their *willingness* to afford job or earnings security. The capacity may be purely financial where labour costs are in competition and the employer needs to transmit the effects of downturns to employees. However, some firms hoard labour in these circumstances while others do not (Bowers *et al.* 1982). Moreover, some firms insulate a core labour force while allowing others to suffer employment insecurity (Sabel 1982). There appears to be some looseness of fit between product-market experience and labour policies. The choices made by firms in the latter area may thus be important determinants of the acceptability of change.

CONCLUSION

In this chapter I have tried to trace the links between company strategy, technological change, and the likelihood of trade union resistance by looking more closely at the nature and incidence of innovation in post-war Britain. Overall, trade union resistance is least likely in industries characterized by rapid product innovation and most likely where cost-cutting process changes prevail. Because of this, the arguments presented by trade unions and discussed in Chapter 2 need to be reconsidered. An overall TUC policy of encouraging innovation and bargaining over its consequences would tend to have an uneven impact: it would perhaps improve the rate of change in performance-maximizing industries but might frustrate innovation and harm competitiveness in mature industries where cost is important. Affiliated unions are likely to be differentially affected: those organizing innovative industries might experience employment expansion on the basis of development of products which become cost-minimizing process innovations in other industries. If the arguments of this chapter are correct, bargaining over technological change is likely to be most difficult where, on the basis of employment and output considerations, unions would see it as most necessary. In particular, the performance-maximizing sector excludes those industries which appear most strike-prone in the face of change.

It is now possible to say rather more about the 'innovation problem' in the UK. It will be recalled from the previous chapter that the UK appears to experience particular problems with incremental changes in large organizations: these sorts of change are characteristic of improvements to process efficiency in the cost-minimizing sector. However, the *overall* rate of innovation also depends rather more on the innovations of the performance-maximizing sector and here, aside from the post-war strike-proneness of engineering which Durcan *et al*. demonstrate, the issue appears to be the absence of innovation rather than opposition to it. It may be, then, that trade unions act as a brake on the efficiency of mature processes rather than an obstacle to the introduction of new products. If this is the case, then the relationship between technical innovation and organizational efficiency needs to be explored in rather more detail. The literature discussed in Chapter 2 suggested some broad relationships between efficiency and the organization of interests at the level of the national economy. However, as I have emphasized in this chapter, innovation occurs within the context of industrial organizations, as does the achievement of efficiency, and some organizations are clearly better than others in both innovative and efficiency respects. In the next chapter I shall explore the factors underlying this phenomenon, with particular reference to the institutions of labour management.

NOTES

1. For discussions of the relative merits of these two general approaches, see Mowery and Rosenberg (1979) and Freeman *et al*. (1982, 35–44).
2. See Abernathy and Townsend (1975, 381): 'the unit of analysis is defined to include the physical process (capital equipment, tasks, labor skills and process flow configuration), the product, the characteristics of input materials and the characteristics of product demand that are incident on the process' (*sic*).
3. This relies heavily on Utterback (1979) and Utterback and Abernathy (1975).

4. Adaptation is according to Central Statistical Office conversion guidelines.
5. The means for the 'non-engineering' performance-maximizers in Table 4.4 are: employment growth 0.22; growth of output per head 5.71; growth of unit wage costs 1.56 (Figures are 'per cent per annum').
6. Townsend *et al*. (1981) and Pavitt (1984) show a relatively low level of innovative activity in the UK sales-maximizing MLHs, although a fairly large number of the innovations are process changes. However, the SPRU data under-represents the cost-minimizing sector in omitting transport and communications, and the sales-maximizing sector in omitting brewing, tobacco, and services.
7. The average for 1949–73 is 63 working days lost per 1000 workers employed in chemicals compared with 369 in electrical and non-electrical engineering (Durcan *et al*. 1983, 176).
8. Not necessarily continuously: economies of scale become most important during the middle of the sales-maximization period.
9. There may be productivity consequences as well. Abernathy and Townsend note that strong relationships between R&D expenditure and productivity improvement apply in the USA only to science-based second-stage industries (1975, 381).

5

An Approach to Institutional Change

INTRODUCTION

THE purpose of this chapter is to explain why some sets of industrial relations institutions appear to be better at accommodating change than others; the approach developed here will guide the analysis of those industries that have experienced most resistance to change which will be presented in the next two chapters. I shall argue that these industries show similar institutional arrangements for the management of labour. The chapter will focus on the impact of technical change on both procedural and substantive issues, but I shall depart from conventional industrial relations in two respects: first, I am concerned not merely with collective bargaining, but also with the organization of work and with the impact of technical change on individual effort bargains; secondly, I shall in consequence employ a rather different theoretical approach which has its roots in institutional economics rather than sociology. Although I shall draw a number of parallels between these two standpoints, the basis of the approach to change presented here is the analysis of idiosyncratic exchange.

The structure of the chapter is as follows. Initially, I shall discuss different forms of procedures for governing employment relations over time. Subsequently, I shall discuss the substantive impact of new technology on effort bargains and work organization. This division is primarily for presentational purposes, since in practice procedural and substantive issues are unlikely to be independent. I shall begin with a brief description of the theoretical background.

THE ANALYSIS OF EMPLOYMENT CONTRACTS

One of the basic concerns of institutional economics is that a large class of transactions which are of interest to economists takes place within organizations rather than in markets; moreover, many transactions tend to be recurrent and long-term, and to involve small numbers on each side acting under conditions of uncertainty. In such conditions, exchange becomes 'idiosyncratic' in the sense that neither party can easily go elsewhere for a similar transaction without incurring substantial costs. From this insight emerges a concern with the most efficient form of transaction where recurrent contracting occurs. At base, this concerns the division of economic activity between the firm and the market, but extends to a discussion of the development of different institutional forms within enterprises for the most efficient conduct of different sets of transactions. One of the primary focuses of interest is thus the relative efficiency of different forms of organization; in this sense, institutional economics is, crudely, concerned with x-efficiency (Leibenstein 1976).

The relevance of this line of reasoning for industrial relations follows from Doeringer and Piore's recognition that the employment relation is characterized by idiosyncratic exchange:

> Almost every job involves some specific skills, even the simplest custodial tasks are facilitated by familiarity with the physical environment specific to the workplace in which they are being performed . . . (1971, 15–16.)

In fact, task idiosyncracies are of several sorts. In Williamson's view they involve:

(i) equipment idiosyncracies, due to incompletely standardized equipment, the properties of which become known only through work experience;
(ii) process idiosyncracies, adapted or fashioned by employees themselves in particular operating contexts;
(iii) informal team accommodations, attributable to interpersonal adaptation, which do not survive personnel changes;
(iv) idiosyncratic information demands and codes specific to particular organizations. (1975, 62.)

Where idiosyncratic exchange occurs, a rather different theoretical approach from that of conventional economics is required. Williamson's approach—the so-called 'organizational failures' framework—is presented in Figure 5.1. He argues that idiosyncratic exchange is characterized by the low number of alternative exchange opportunities available for the parties, that actors have severely 'bounded' rationality, that both sides of the bargaining process are prone to act in 'opportunistic' (i.e. strategically dishonest) fashion where circumstances permit, and that information is 'impacted' (i.e. unevenly distributed). The model applies to the employment rela-

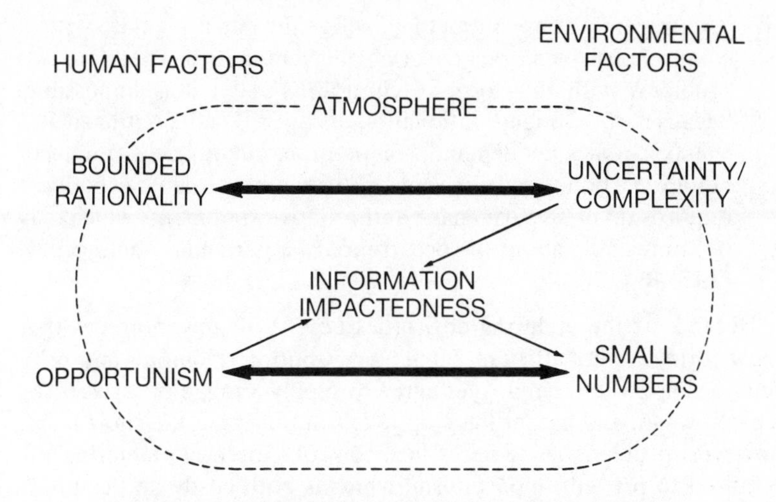

Figure 5.1. The Organizational Failures Framework
Source: Williamson 1975, 40

tionship where the task idiosyncrasies listed above imply that workers acquire job-specific skills and knowledge.

In analysing employment contracts, the problem Williamson poses is as follows: how does the buyer of labour secure efficient labour contracting given that suppliers (job-incumbents) are prone to opportunistic behaviour and possess strategic advantages over outsiders in competitive bidding? The potential for loss follows from a tendency of employees themselves to realize the advantages of task idiosyncrasy, preventing the benefits of such efficiency improvement from flowing to the firm. Williamson thus seeks to outline the advantages and disadvantages of different forms of contracting for labour, and to interpret 'evolving institutional practices with respect to idiosyncratic production tasks, principally in efficiency terms' (1975, 57).

The main thrust is to show the problems involved in two different forms of contracting. These are:

(i) 'Sequential spot contracting', where the employer contracts with the employee to provide a service as the demand for it arises. The principal problem here is that the employee will act opportunistically, and bargain a high price for the service since he or she operates from a position of strategic advantage secured by job idiosyncrasy.

(ii) 'Contingent claims contracting', where the parties seek to write a comprehensive contract covering all eventualities. There are two problems with this form of contract. Firstly, it is impossible, because of bounded rationality, to specify all contingencies: changes in product demand, competition, factor prices, and technology are bound to be in some measure unforeseen. Secondly, it requires that one party make authoritative statements which may be contestable about the occurrence of a particular contingency. (1975, 64–70.)

Because of the enduring difficulties here, Williamson argues that these forms are not efficient.[1] One further option is Simon's authority relationship, where employees agree to supply a range of services for a given wage, and accept the employer's authority in the short term; however, problems arise here since control ('intensive metering') is required to prevent opportunism, which is both costly and counterproductive. The most efficient form of labour contracting is essentially Doeringer and Piore's idea of an internal labour-market. Williamson emphasizes four elements:

(i) Wage rates are attached to jobs rather than workers; incentives for opportunism typical of individual bargaining are attenuated, and group disciplinary disincentives are also deployed.
(ii) Agreements are reached through collective bargaining, and unforeseeable contingencies are referred to arbitration machinery.
(iii) Grievance procedures allow changing conditions to be dealt with in a systematic way.
(iv) Internal promotion ladders are established which encourage a positive worker attitude towards on-the-job training and enable the firm to reward co-operative behaviour.

The essence of the comparative advantage of internal labour-markets is the securing of x-efficiency gains through an impact on employee attitudes. Close supervision, which can have de-motivating consequences, is avoided.[2] As a consequence, a form of involvement termed 'consummate co-operation', which has beneficial efficiency properties, is encouraged. The distinction from 'perfunctory co-operation' is as follows:

Consummate co-operation is an affirmative job attitude—to include the use of judgement, filling gaps, and taking initiative in an instrumental way. Perfunctory co-operation, by contrast, involves job performance of a minimally acceptable sort—where minimally acceptable means that incumbents, who through experience have acquired task-specific skills, need merely to maintain a slight margin over the best available inexperienced candidate. (Williamson 1975, 69.)

The internal labour-market solution thus encourages employees to secure for the *firm* the x-efficiency gains available from job idiosyncrasy. One has, in the emphasis on the accrual of efficiency gains over time, a micro-economic account of the problem of distributional groups outlined in Chapter 2, and some suggestions for its resolution.

Williamson's work thus overlaps at a number of points with other approaches which may be more familiar to industrial relations analysts. In particular, it shares theoretical roots in the idiosyncratic-exchange literature with the so-called 'Harvard' school, who have sought to argue that a *positive* relationship between unionization and productivity may occur. The Harvard school transfer Hirschmann's (1970) insight that those who are displeased with the operation of an organization may leave (exit) or protest (voice) from a product-market to a labour-market setting. At its simplest, the argument is that unionization provides a communications device of greater sophistication than the quit mechanism (i.e. the exit option): since costs attach to search, training, and lost production through quits, unionization (the voice mechanism) can contribute to efficiency gains. Union activity thus parallels the effect of consumer associations on organizational performance. In the later formulation of Freeman and Medoff (1979), two further avenues for efficiency gains are presented. The first is reminiscent of Williamson. Unions encourage team accommodation and monitor shirking: seniority systems reduce competition and enhance on-the-job training. The second is the more conventional 'shock' effect in which the union wage impact encourages the reduction of slack.

The central similarity between Williamson and the Harvard school is the shared emphasis that collective bargaining may actually facilitate change and the improvement of efficiency. Internal labour-markets for Williamson and 'voice' for the Harvard school are both institutional arrangements which help the organization to accommodate change. However, imperfect contractual arrangements or the 'other face' of trade-unionism (for Freeman and Medoff) may also lead to inefficiency. The net effect of unionization is an empirical question—a point which several critics lose sight of (Addison and Barnett 1982; Addison 1983).

In addition, Williamson's use of the distinction between 'perfunctory' and 'consummate' co-operation is simply shorthand for ideas which have been more fully developed in sociological literature. For example, Fox's (1974) discussion of high- and low-trust relations links employee attitudes, contractual form, and work organization in a similar way, although without an economic analysis of exchange or efficiency. Similarly, Offe (1976) makes a distinction between the 'initiating' and 'preventative' impacts of employees on poduction processes.

I shall have more to say about these comparisons below. The advantage of the organizational failures framework is that, as I hope to show, it allows quite specific predictions about the impact of specific forms of change on specific forms of institution. Moreover, its relevance for the present purpose is enhanced by the view taken of technological change. Idiosyncrasies stemming from technological considerations do not exhaust the range of task idiosyncrasies. Williamson has argued at some length that the growth of industrial organization follows from transactional rather than technological considerations and, indeed, that certain types of technology may obstruct transactional efficiency (1975, 61–5; 1980; 1982). Technological choice follows from efficiency considerations and

> Although least-cost production technologies are sacrificed in the process, pecuniary gains may nevertheless result since incumbents realise little strategic advantage over alternative qualified but inexperienced outsiders. (1975, 68.)[3]

Technological change is thus set firmly within the context of corporate strategy: in fact, organizational innovations such as the development of divisional organization are considered as equally important sources of competitive advantage. Williamson tends to discuss efficiency rather than innovativeness; however, it is central to an understanding of his contractual approach that efficient performance depends upon successive adaptations to incremental changes in process and product. The conflict noted by Abernathy *et al*. (1983) also emerges, and Williamson seeks to propose remedies to the perceived inability of bureaucratic structures to generate radical change (1975, 132–55, 63–4, 192–207).

AN AMENDED APPROACH[4]

The organizational failures framework has been subjected to several criticisms. In its application to employment relations, the most obvious point to make about Williamson's approach, is that it ignores issues of power (Marglin 1976, Francis *et al.* 1983), and indeed Williamson does regard power as a residual category.[5] I have previously made some rather more specific points about the sociological basis of the ideas about co-operation and the validity of Williamson's views on internal labour-markets (Willman 1982b, 1983a): I shall return to these below. However, these criticisms notwithstanding, institutional economics provides a useful framework for the analysis of collective bargaining over change.

In Figure 5.1, opportunism emerges as a potential form of behaviour for all of those who have the advantage of small numbers bargaining. However, in Williamson's discussion of the employment relation it appears only as a feature of *seller* behaviour; buyers are seen predominantly as boundedly rational. However, where buyers distribute sellers' products in a market to which sellers themselves do not have access, or where sellers are unaware—through information impactedness or bounded rationality—of the nature of final product demand, buyers may be opportunistic and the boundedness of sellers' rationality becomes important.

I simply wish to suggest here that the 'human factors' *do* apply universally to buyers and sellers. This possibility is recognized by Williamson at one point, but does not feature significantly in his discussion. The employer, he says, will *not* behave opportunistically

> *unless*, at the time he got the worker to agree to a wage w, he represented to the worker that services of type x_i would be called for when event e_i obtained when in fact x_i' services, which the worker dislikes, yield a greater e_i gain. The worker, being assured that he would be called on to perform x_i' services only when the unlikely event e_i' occurred, agreed to a lower wage than he would have if he realized that an x_i' response would be called for in both e_i and e_i' situations—because the employer will falsely declare e_i to be e_i' so as to get x_i' performed. (1975, 66.)

This is potentially a large class of events. On the assumptions of the framework that all social actors are prone to opportunistic behaviour, that employers are faced with uncertainty/complexity, and that their rationality is bounded, opportunistic behaviour is plausibly a prevalent form of self-insurance. Let us assume further that the small-numbers situation confronts the employee and the environmental conditions are fulfilled: typically, employers will have a great deal more information about the company of direct relevance to the employment relation than will employees. Such information may cover changes in labour costs as a proportion of total costs, changes in the value of the contribution of different categories of labour, or changes in production technology which may involve labour shedding.

Employers may use privileged information about product-market conditions and the state of the firm to consistently misrepresent the real world, maximizing their own scope for opportunism. They may similarly effect organizational changes to implement internal labour-market structures in order to minimize the scope for worker opportunism. Workers' strategies cannot usefully be seen as a simple mirror image of this, primarily because they cannot normally reorganize the firm. They may try to introduce uncertainty into the contractual relationship for their own advantage: for example, by pressing for sequential spot contracting. They may try to restrict the employer's capacity for opportunism by constraining him within a complex contingent claims contract, but, in doing so, they may deny themselves manœuvring space.

The implications of this for Williamson's analysis are clear: the internal labour-market emerges as an efficient solution to the transaction-cost problem at least partly because of its 'atmospheric' properties. Admit bilateral opportunism, and these properties may evaporate. For internal labour-markets to be an acceptably neutral resolution of the problem of bilateral opportunism they would need to constrain the opportunistic tendencies of both sides in equal measure, but they do not. The standardization of rules in an internal labour-market and the existence of a grievance procedure may prevent what one might term '*capricious*' or discriminatory opportunism, directed at particular employees within the work-force. But employers are still free to be opportunistic in their use of heavily impacted information about state-of-the-world conditions, and may persist in *systematic* opportunism, directed consistently at the sale of labour as a whole: this opportunism remains unconstrained.

It follows, then, that internal labour-markets would not of themselves guarantee the minimization of transaction costs, primarily because their alleged atmospheric benefits are unlikely to be forthcoming. They constrain the behaviour of the two parties unevenly:

the form is in fact a sophisticated contingent claims contract. Once this is recognized, two questions arise. Firstly, is it possible to specify an appropriate institutional solution to the problem of opportunism within the organizational failures framework which would be acceptable to both parties and, if so, under what circumstances? Secondly, what might one expect about the contract preferences of employers and workers acting under conditions of bilateral opportunism where such an institutional solution is unavailable?

The starting point here is a formalization of Williamson's discussion of contractual forms (1975, 57–82). Under sequential spot contracting, the employer (B) and the worker (W) agree a wage (w) for the delivery of a service (X) as and when the demand for that service arises. Under contingent claims contracting, B contracts with W for the delivery of services $x_i, \ldots, x_n$ in return for wages $w_i, \ldots, w_n$ contingent upon events $e_i, \ldots, e_n$ occurring in the future. These two types are represented in Table 5.1.

In addition, Williamson also discusses Simon's 'authority relationship', from which the notation above is adapted. In this framework B selects the X he wishes W to follow. X can be differentiated into subsets $x_i, \ldots, x_n$, only some of which will be within W's 'area of acceptance'. The employment relationship will be preferred when

> It does not matter to him (i.e. W) 'very much' which X (within the agreed upon area of acceptance) B will choose, or if W is compensated in some way for the possibility that B will choose X that is not desired by W (i.e. that B will ask W to perform an unpleasant task). (Williamson 1975, 69.)

As we have seen, all three types are flawed. Sequential spot contracting is chronically prone to opportunism, contingent claims contract-

Table 5.1. Sequential Spot Contracting and Contingent Claims Contracting

Sequential spot contracting: time, $t_i, \ldots, t_n$				
Range of agreement	t_i	t_j	t_k	$t_i, \ldots, t_n$
	e_i	e_j	e_k	$e_i, \ldots, e_n$
	x_i	x_j	x_k	$x_i, \ldots, x_n$
	w_i	w_j	w_k	$w_i, \ldots, w_n$
Contingent claims contracting: time, t_i				
Range of agreement	$e_i, \ldots, e_n$			
	$x_i, \ldots, x_n$			
	$w_i, \ldots, w_n$			

ing fails to economize on bounded rationality, while Simon's model fails to consider efficiency aspects and is prone to opportunism in the absence of close supervision.

It seems clear from this formalization that internal labour-markets will not successfully attenuate opportunism. They would provide the mechanism under which employers could meter the provision of $x_i, \ldots, x_n$ and, in the standardization of rules, they would introduce a form of equity. However, they provide no basis for employees to monitor $e_i, \ldots, e_n$ or $x_i, \ldots, x_n$ in relation to $w_i, \ldots, w_n$. Specifically, the internal labour-market mode does not deal with two problems of asymmetric (or 'impacted') information.

(i) concerning the occurrence of $e_i, \ldots, e_n$; and
(ii) concerning the relationship between the values of $w_i, \ldots, w_n$ and of the $x_i, \ldots, x_n$ for which they are provided;

nor with two problems of organizational control:

(iii) the control the employer has over the occurrence of $e_i, \ldots, e_n$ directly by decision or indirectly by exerting influence on the environment of the firm;
(iv) his ability to specify services $x_i, \ldots, x_n$ and to change the demand for such services.

Thus, internal labour-markets may provide a means for establishing predictable relationships and a form of equity between employees, but on their own they do very little to guarantee equity in the effort bargain; under these circumstances, it is unlikely that internal labour-markets would invariably generate consummate co-operation. If employees are to forgo opportunism rather than be disciplined into avoiding it, then, within the logic of the model, modifications are necessary both to the degree of information impactedness and to the hierarchy of decision making.

1. *Information impactedness*. The principle areas of uncertainty for the employer concern the provision of the range of services $x_i, \ldots, x_n$ by the employee. The techniques for the removal of such uncertainty consists of a mix of:

(a) supervision;
(b) the measurement of work and of work intensity by techniques of time study and effort rating;
(c) the construction of jobs in such a way that they consist of known predictable elements (the basis of this is method study).

It is thus possible for the employer to use such techniques to secure information about direct inputs to the provision of services: this deals effectively with a range of equipment and process idiosyncrasies. Team accommodations and communication channels may be further adapted by techniques of job design or team building. However, the clear risk in the use of these various techniques is that consummate co-operation may be withdrawn, and efficiency losses experienced.

By contrast, the principal areas of uncertainty for workers concern the relative values of $w_i, \ldots, w_n$ and $x_i, \ldots, x_n$: in these areas, the mechanisms for obtaining information are much less developed. Legal provisions for disclosure of information, including those specifically designed for use in collective bargaining, are unlikely to provide adequate information (Gospel and Willman 1981); informal, clandestine, or illicit sources may be of little use. However, unless substantial voluntary disclosure of such information occurs, it follows that consummate co-operation will be jeopardized. The removal of opportunistic behaviour by employees depends upon the employer's own ability to disarm *their* expectations that he, too, will behave opportunistically given the chance.

2. *Decision making*. If workers are to receive payment for the provision of $x_i, \ldots, x_n$ consequent upon $e_i, \ldots, e_n$, they may be presumed to require influence over their selection or occurrence (in so far as this is possible) in return for consummate co-operation. The firm's influence over $e_i, \ldots, e_n$ may only be significant where it has considerable influence on its market environment. Of more importance is the set $x_i, \ldots, x_n$. Williamson explicitly allows that the services required by the employer will follow from employer decisions on transactional factors rather than simply from technological considerations. In an important sense, then, the set $x_i, \ldots, x_n$ is indeterminate. In return for consummate co-operation, employees may require influence over the work-organizational features which delineate $x_i, \ldots, x_n$ or at least assurances about the construction of the set.

This raises a number of efficiency issues. Although hierarchical decision making allows employer opportunism, workers may prefer it to the extent that it seems to be more efficient than more egalitarian forms. However, it should be stressed that hierarchical modes other than authority or internal labour-market ones exist, and that these do not involve the same degree of information impactedness. For example, collective forms involving rotational leadership may be only marginally less efficient than authority relations, and may be superior in terms of atmosphere (Williamson 1980, 21–30); hierarchy persists but information impactedness is much less severe. It is possible, then, that workers might prefer peer-group structures even where efficiency concerns are uppermost, since the threat of opportunism is removed. However, on the assumption that peer-group organization is not feasible in large industrial organizations, one returns to the conclusion that internal labour-market structures need not only to be supported by good disclosure practice but also that decisions must be open to practical modification through employee influence.

There are thus four necessary conditions for the avoidance of opportunism over time: equity between employees, mechanisms of adjustment to product-market change which offer security, adequate disclosure of information, and access to decision making. If any of these conditions is absent, then one might expect employers and employees to be concerned about opportunism in their choice of contractual form. This contention is central to the amended framework offered here.

This framework runs as follows. Assuming that the environmental conditions of the organizational failures framework hold and, further, that there is some form of collective organization on the part of the work-force, then employers and workers, both confronted by uncertainty, will seek in their relations each to reduce the amount of uncertainty with which they are confronted, consistent with the 'atmospheric' requirements of remaining on sufficiently good terms for the relationship to persist. This pursuit will influence the contractual arrangements preferred by each. Given the employer's selection of contractual arrangement, workers reactions to undesirable forms will consist of behaviour designed to modify the conditions of contract in the preferred direction.

As an illustration, consider the following two models of collective bargaining. In the first, the parties arrive at a substantive agreement on all matters subject to joint regulation. This agreement will have a fixed term, during the course of which industrial action is not permitted, recourse being to a procedure for *interpretation* rather than extension of the terms of the initial agreement. In the second, overall agreement is restricted to a disputes procedure to resolve differences: particular substantive rules have no fixed terms and may be negotiated when change occurs. The union is free to take industrial action on procedural exhaustion, and no clear distinction exists between disputes of right under existing agreements and disputes of interest over the terms of a new agreement (Clegg 1979, 116–17).

In important respects, these two forms of agreement are similar to those required by Williamson's contingent claims and sequential spot forms respectively. In the first instance, the major problem in compiling the agreement is to generate consistent and comprehensive coverage of contingencies to enable the procedure to function as an interpretive rather than as a generative device. In the second, opportunism is the main problem: given that jobs are idiosyncratic and that the employee may also be confronted by a small-numbers situation, 'moral hazard' is endemic. The adoption of either form of agreement in a given circumstance is obviously to some extent indeterminate, depending upon the preferences of the parties. However, their likely preferences can be spelt out: indeed, some indications of what is involved have already been given.

The first model, where the main contingencies are adequately catered for, limits the scope for opportunism on both sides. However, this limitation may not be symmetrical: enforcement problems remain. Asymmetrical information is likely to arise since the introduction of new work items and decisions about the product-market are generally made by the employer. In practice, employers may thus have less bounded rationality, and the threat of opportunism perceived by the work-force may be more severe, under contingent claims arrangements.

The second model allows both parties considerable scope for opportunism. Where, in the medium to long term, conditions favour one party to the consistent disadvantage of the other, opportunistic success will be directly related to the frequency of bargaining. For example, in the post-war period in Britain, tight labour-markets and a well-developed range of shop-floor bargaining tactics allowed wage increases to be frequently exacted, often, it is claimed, without commensurate productivity increases. By contrast, in the past, rate cutting occurred where the small-numbers condition was eroded by high unemployment. The preference for sequential spot bargaining is thus straightforwardly explicable in terms of the perception of labour- and product-market condition and the pursuit of short-term self-interest.

Defensive strategy involves the selection of those forms which minimize the opponent's scope for opportunism. Offensive strategy seeks forms which maximize one's own opportunistic scope. The significance of this argument lies in its focus on the contestable nature of contract preference. However, given the empirical infrequency of a change in contracting modes in industry, the tactics of workers' opportunistic behaviour—both offensive and defensive—under a particular regime also constitute conditions of interest.

In a tight labour-market, or a buoyant product-market, the offensive strategy of maximization of opportunism will prevail on the worker side, and the defensive strategy of preventing this will prevail for the employer. Particularly if the trend appears likely to persist for some time, employees will prefer frequent renegotiation of terms at successively more advantageous rates, while employers will seek to stabilize the situation. The appropriate contractual preferences are thus, for the employees, sequential spot contracts and, for the employer, contingent claims contracts. In a slack labour-market, or one where labour is becoming more easily available and its costs are falling, the defensive stragey of minimizing employer opportunism will prevail with employees, and the offensive strategy with employers. With a persistent trend, employees will prefer contingent claims contracts to sequential spot ones. Employers, by contrast, may well prefer sequential spot contracts. The framework implies a continuous tendency for the decay of contingent claims contracts both because of the occurrence of unforeseen contingencies and because of unilateral pressure in favour of sequential spot contracts under either regime. However, asymmetry of control is important. The employer is concerned to encourage consummate co-operation. Recourse to successive spot contracting may be tempered by a concern with its effects on employee morale through the introduction of uncertainty. Consequently, the employer may move to an authority relationship, where a range of services are specified for a given wage, reducing employee uncertainty about the range of services required without the requirement to reach agreement on a wage for each of them. The preferences of the different parties to contracts under different circumstances may thus be as in Figure 5.2. This logic may now be applied to discussions of technical change: this requires that it be related to the discussion of previous chapters.

PROCEDURAL ISSUES: TECHNICAL CHANGE AND COLLECTIVE BARGAINING

Within the organizational failures framework, the organization of work and, consequently, the set of services $x_i, \ldots, x_n$ for which wage

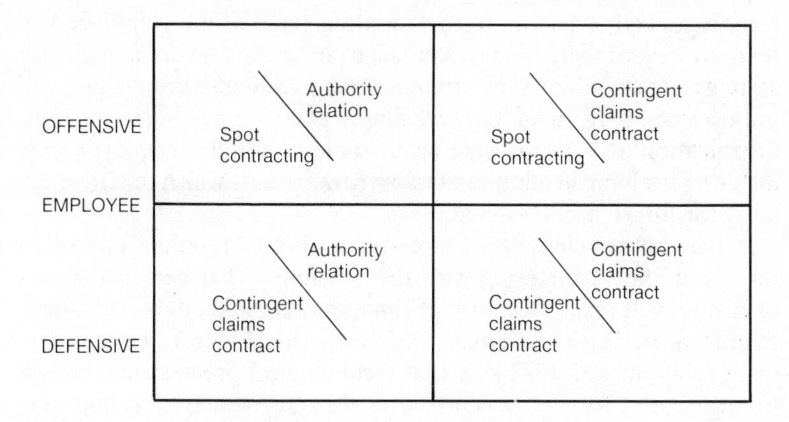

Figure 5.2. Agreement Preferences

rates $w_i, \ldots, w_n$ will be paid do not follow directly from the adoption of a particular technology: decisions which concern them may be negotiable. Technical change tends to generate information asymmetry in favour of the employer who typically indulges in R&D or market search before installation of equipment which will affect work organization: it thus opens the way for opportunism.

There are, thus, in terms of the reasoning so far, three sets of issues around which negotiations can take place under conditions of technological change:

(i) The event e_i (i.e. the decision whether or not to install new equipment).
(ii) The service x_i (i.e. the organization of work associated with the new technology).
(iii) The wage w_i (i.e. the wages to be paid for the work involved).

Clearly, changes which are associated with unfavourable wage consequences or which are seen to have unfavourable consequences for work organization would, a priori, be unwelcome, and this would affect decisions over purchase: this will be discussed below. However, different contractual arrangements are likely to respond to change in different ways.

1. *Authority relationships*. Where labour-markets are slack, it might be expected that both the employer and the employee will

sustain a large set $x_i, \ldots, x_n$ for a specific w_i. The employer will not foresee a need to pay more to widen the set, and the employee will tend to regard bargaining over such an extension as a high-risk strategy. Thus, under such circumstances, technological change will not occasion bargaining, but will simply generate new forms of work organization at a given wage level. Benefits for the employee may include retraining in additional idiosyncratic skills which might be the basis for future earnings increase.

It will be recalled that successful authority relationships often require intensive metering with the possibility that perfunctory co-operation will result from use of time- and method-study techniques to enforce it. This is essentially the 'Braverman' view of the employment relation, and it is likely that technological process changes will be introduced for the reasons he envisaged: namely, the improvement of control, the cutting of costs, and the divorce of conception from execution (Braverman 1974). In terms of the classification of Chapter 4, authority relationships may be the preferred employer strategy when undertaking innovation for cost-minimization purposes.

2. *Sequential spot contracting*. This form is, on the face of it, well adapted to recurrent product and process change where sufficient uncertainty exists to prevent the drafting of comprehensive contracts: employers are particularly likely to favour it where product-market conditions are variable, where labour-markets are slack, or where labour costs are relatively unimportant. It is thus a likely form for performance-maximizing industries, particularly batch production where relatively inefficient production systems are not unduly damaged by recurrent bargaining.

However, under certain conditions it allows employees considerable control over the terms of provision of $x_i, \ldots, x_n$: in the economist's terms, it promotes the growth of restrictive practices. It is thus wholly inappropriate for production in sales-maximizing industries or cost-minimizing ones where economies of scale are important. Efficient production systems based on low product diversity and high levels of automation are likely to be disrupted by recurrent bargaining.

3. *Contingent claims/internal labour-markets*. Both forms are appropriate for predictable 'systemic' production systems in which innovation may occur but is likely to be of the incremental variety.

However, in reducing uncertainty, they reduce opportunistic scope and the imminence of larger-scale changes, which cannot be specified in a comprehensive contract or which so change the set of services required that internal labour-markets cannot function, may cause the movements outlined in Figure 5.2: employers seek not to negotiate, and employees favour spot bargaining.

Different forms of contract are thus unevenly receptive of change. However, it is important to emphasize that none of these contractual modes will be stable. The authority relationship and sequential spot contracting are both vulnerable to shifts in strategy under changing market conditions. Complex contingent claims contracts suffer from bounded rationality constraints: i.e. they will be subject to supplementation by authority relationships or spot contracting in the event of unforeseen contingencies. The stable contractual mode for the negotiation of technological change will reduce information impactedness while granting employee involvement in decision making.

This sheds new light on recent initiatives for new technology agreements, and on the differences between the UK and USA discussed in Chapter 2. Whereas the TUC check-list stresses influence over design of equipment, change only by agreement, consultation, and advance notification, management-rights clauses in the USA tend to restrict trade union contract language to the latter two items (Murphy 1981).

Contract provisions developed in the first automation debate in the USA tended to be complex contingent claims contracts which specified employer commitments in the event of changes to seniority, arrangements, pay, relocation, or severance (Blitz 1969). An addition to this, to reduce information impactedness, is the provision of a clause on advance notification: by 1978, 11 per cent of agreements contained such clauses (BLS 1980). Further developments to influence decision making directly are rarer. Model contract language developed by the machinists' union proposed the establishment of

> a Joint Union Management Committee for Technological change, composed of equal representation from the Union and Management, to study the problems arising from technological change in relation to the effect on employees in the bargaining unit. (IAM 1980.)

However, this has yet to receive widespread acceptance by incorporation into agreements.

Similar problems existed in the UK (Williams and Moseley 1982; Willman 1983b), but it is worth emphasizing that the approach to new technology was not uniformly that of the TUC. The print unions, particularly the NGA, preferred to continue spot contracting from a position of strength in national newspapers (see Chapter 6) while ASTMS developed a draft agreement which largely sought to extend normal collective bargaining to cover new technology: it was thus essentially a spot-contracting form, suggesting that either

(i) agreements previously made would be reopened to take into account the changed work circumstances, or
(ii) the introduction of the new technology would be delayed until bargaining on the appropriate matters had taken place upon the expiry of the existing agreement(s).

Empirically, new technology agreements are probably of minor importance in the regulation of technological change under collective bargaining. The issue, then, is whether a contingent-claims, internal labour-market, or spot-contracting form operates.

The distribution of these contractual forms across the economy is thus of considerable importance. Most large companies, of course, operate with a mix of contractual forms: in the UK, internal labour-markets are more common for white-collar and managerial staff, and authority relationships more common for non-unionized employees. However, there are sectoral patterns: for example, comprehensive contracting associated with the development of internal labour-markets is typical of much of the unionized financial sector and the public sector. Sequential spot contracting exists in printing, dockwork, and those parts of the engineering industry covered by agreements on mutality. In part, this relates to product-market differences. For example, the commitments involved in developing internal labour-markets do tend to turn labour costs into fixed costs, and the form is most appropriate where demand for a product sustains steady demand for at least part of the labour-force. Sequential spot markets tend to be associated with industries where product-market fluctuations have produced variable employment levels.

This latter correspondence is of considerable interest. Slichter *et al.*, focusing on the USA, found that variable product-markets were associated with the development of make-work rules and restrictive practices: this was particularly likely to be the case where markets were local and the product perishable (1960, 335–6). For the UK, we have shown above that those industries which accounted for most industrial conflict over technological change were historically associated with spot contracting: docks, printing, and motor vehicles. More precisely they were, during the period under discussion, in the process of transferring to forms of labour contracting in which bargaining was restricted. These industries will be the subjects of further discussions in Chapters 6 and 7. However, having discussed the influence of technological changes on contractual forms, I now turn to discussions of substance.

SUBSTANTIVE ISSUES: EFFORT BARGAINS AND WORK ORGANIZATION

If new technology is associated with increased intensity of work, it may be resisted. The discussion is assisted by Figure 5.3, derived from Baldamus (1961). Along the line AB, effort and wages are assumed to be in balance; movements either up or down the line maintain this balance. Within a firm, the spread of effort levels may be BC, which will also represent the spread of wages. The line AB runs at an angle of 45 °.[6] Several factors may change the slope. All forces which improve the employer's bargaining position will tend to shift the line clockwise; those which improve the employee position will tend to shift it anti-clockwise. Technical change which tends to improve productivity at constant effort may promote anti-clockwise movement; to the extent that it destroys bargaining power, however, it will have the opposite effect. Movements along AB may also be problematic. If

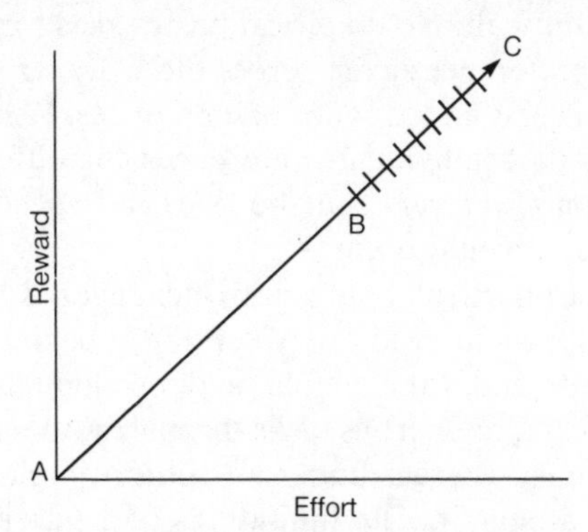

Figure 5.3. The Effort–Reward Curve
Source: Baldamus 1961

technical change moves individuals up or down the line, problems may arise. For example, continuous shift working may be introduced which raises effort and reward to maximize the utilization of capital equipment; or employees may be asked to work part time if capital is substantiated for labour. Neither of these will necessarily be acceptable to employees.

A number of distributional issues are involved in concerns over work organization. Individual effort-bargain adjustments require complementary changes where work activity is co-ordinated. Certain individuals may free-ride on the effects of change; or the effort-value rank-ordering BC may be disturbed, causing resistance to change. In addition, as Leibenstein notes (1976, 95–134), changes to the effort position of an individual or group are only likely to be voluntarily effected when the utility gains exceed the costs of change: the latter will include the costs of co-ordinating a new organization of work.

Job-content issues further complicate matters. Occupational groups tend to have reasonably consistent views about the relative worth not only of occupations but also of job elements (Willman 1982a). To use the notation already introduced, within groups there may be a stable preference ordering for a set of tasks $x_i, \ldots, x_n$. Technology which removes the need for x_i, a low-preference task, may be acceptable, but one which automates x_k, a high-preference one, may be rejected. Similarly, technology which removes the need for idiosyncratic tasks may be seen, independent of job preferences, to erode bargaining power and hence the basis of privileged exchange. Further generalizations about the impact of technical change are thus difficult. No empirical basis exists for the suggestion that stable task preferences exist across the working population as a whole, nor even across large groups of employees. Williamson (1975) himself suggests that individual preferences for different types and mix of transactions will vary, but he does not specify the basis on which the variation would occur.

The response to particular substantive changes is thus difficult to predict. There is a relationship between the substantive change and the procedures adopted: for example, under sequential spot contracting the removal even of onerous tasks through automation is likely to occasion bargaining, whereas under an authority relationship it may be perceived as a benefit. The limitations of a purely economic approach are revealed as one moves from the general to the particular.

LIMITATIONS

The transactional approach pursued by Williamson produces some unlikely bedfellows: the view that industrial organization depends upon considerations of efficiency rather than technology and that technological change is subordinate to economic requirements is very similar to the radical views promoted by Marglin and by Gordon.[7] The difference is essentially in the identification of culprits: for Marglin opportunism is characterstic of capitalists, whereas for Williamson, it is a feature of sellers of labour power (Marglin 1976; Gordon 1976; Willman 1983a). I hope that the analysis above illustrates that neither generalization need follow from the organizational failures framework.

However, one area in which differences do exist concerns the operation of power. In the work of Marglin and Gordon, and indeed in that of the 'labour process' writers of the post-Braverman period, power, or at least the related idea of control, occupies a central place in the analysis of employment relations. The idea of control is, of course, embodied to some extent in the form of contractual arrangement, but ideas which relate to industrial discipline are muted in this presentation in part because, bounded rationality aside, the assumptions about individual behaviour are economically quite conventional: hence the view that

> The best evidence that power is driving organisational outcomes would be a demonstration that less efficient modes that serve to concentrate power displace more efficient modes in which power is more evenly distributed. (Williamson 1980, 30.)

However, as several authors, most notably Turk, have pointed out, Williamson is operating with a very specific definition of efficiency in presenting empirical justification for such a view. Not only is the role of power in setting the parameters of economic activity relatively muted but, on a different definition of efficiency, the superior efficiency properties of successive institutional forms are less obvious (Turk 1983, 191–3; McGuinness 1983). This is rather more of a problem for the unamended version of the framework; in the amended version, power considerations enter through the contest over contractual form. The view offered here is that efficiency considerations of the form Williamson considers exercise their influence in the longer term. However, power considerations are important in considering how improvements to efficiency overcome the resistance of established groups. The essence of institutional economics is, as Hirschmann (1970) implies, that relatively inefficient organizations can exist for some time, and that some have the capacity to remedy their defects before extinction. Issues of power and control influence the availability of such remedies but, as Littler and Salaman (1982) note, industrial capitalism is concerned with economic success rather than power *per se*. I shall return to these considerations in the concluding chapter.

The second set of issues concerns the use of shorthand terms such as 'atmosphere' and 'consummate co-operation' to describe attitudinal and behavioural patterns which are sociologically quite complex.

The first point to make is, of course, that the term 'consummate co-operation' covers a range of employee behaviour: for example, it is likely to be very different in performance-maximizing industries, where new product development is important, from that required in cost-minimizing areas with systematic production systems, where simple co-operation matters most. In the terms used by Offe, consummate co-operation covers both those situations where employees have *initiatory* influence over production processes, and those where technology has developed systematically and co-operation is revealed by the absence of *preventative* influence, or 'omissive action' (1976, 36).

The second point is that the extra-economic elements of contractual arrangements have been the subject of detailed analysis in Fox's study of trust relations (Table 5.2). In some ways, Fox's extra-economic elements of contract behave similarly to Williamson's 'attitudinal considerations'. So, for example, one element in the 'high-discretion syndrome' associated with high-trust roles is an absence of close supervision or detailed regulation, and high-trust reciprocation by the occupants of high-trust work roles is created by promoting a given individual through a series of posts each vested with greater discretion than the one before. It is usual for an individual moving thus up the hierarchy to display an increasing intensity of commitment to, and identification with, the organization and its imputed interests (Fox 1974, 114).

The importance of Fox's work in this particular context rests on his discussion of trust 'dynamics'. Institutional economics tends to emphasize a one-to-one relationship between structures and

Table 5.2. Fox's High- and Low-discretion Syndromes

Low-discretion syndrome	High-discretion syndrome
1. Incumbent subjected to close supervision, hedged by impersonal rules	1. Close supervision seen as inappropriate
2. Incumbent perceives supervision as lack of trust in him	2. Role-occupants have moral commitment to organization
3. Close co-ordination of work roles established by routines	3. Problem-solving relations, not externally imposed routines
4. Inadequacies of incumbant seen to result from carelessness or indifference/insubordination	4. Inadequacies due to substandard exercise of discretion
5. Conflict handled through collective bargaining	5. Conflict resolution a problem-solving exercise

Source: Fox 1974.

attitudes, and thus tends to see x-efficiency gains as following from particular institutional forms: hence atmospheric benefits allegedly 'flow' from internal labour-markets. However, the more detailed sociological approach emphasizes the reasons why one might expect conflict in the transfer between one form of employment relation and another.

Fox notes that many 'low-trust' responses have efficiency implications:

> such forms of behaviour as indifferent performance, clock-watching and high absence, sickness, wastage or turnover rates . . . may in varying degrees symptomize the individual's inability to see himself as participating in a fellowship of common purpose and shared endeavour. (1974, 102.)

However, it does not necessarily follow that organizational innovations with beneficial efficiency effects will be unproblematically adopted: both low- and high-trust organizations experience reinforcing 'spirals' of, respectively, a negative and a positive nature. Reversal of such a spiral in the low-trust context implies that one side, usually management, is prepared to risk an opportunistic response to a high-trust initiative. As Walton and McKersie have noted, defensive low-trust approaches appropriate to zero-sum games frustrate initiatives towards problem-solving activity with positive efficiency consequences (1965, 140–5). Purcell, who has applied Fox's framework to a discussion of collective bargaining structure, feels that such defensive responses frustrate change up to the point at which enterprises face crises prompted by this very inability to respond (1981, 232–8).

This problem applies even where contractual changes may bring benefits. When employment contracts are characterized by suspicious vigilance, employees may actively reject changes which actually offer security or higher-discretion tasks. Suppose that employees subjected to spot contracting in which idiosyncrasies had been deployed for sectional benefit (i.e. restrictive practices have developed) are offered internal labour-market structures (on the Williamson model) in order to secure consummate co-operation and thus x-efficiency gains: the change may enhance job security and satisfaction, and increase earnings through the removal of transactional inefficiencies. Fox's study of trust dynamics suggests that employees may resist change for at least two reasons. In the example he notes, craft-workers who have developed high 'lateral' trust relations at the expense of low 'vertical' ones may not adapt to initiatives which raise the latter at the expense of the former (1974, 79–82). In other circumstances, which are particularly relevant to the relaxation of intensive metering, workers may reject change

> either because they are psychologically structured to need-depending and submissiveness, or because they have been so strongly shaped by their subculture and their own adoptive responses to low discretion work that change is seen as distasteful and threatening. (Fox 1974, 115.)

It is perhaps worth emphasizing that organizational changes which depend in part for their efficiency on their motivational properties are unlikely to succeed if imposed: I shall return to this problem below.

CONCLUSION

The approach offered in this chapter is intended to predict certain dimensions of an organization's response to technological change based on analysis of the way in which it contracts for labour: different types of effort bargains are implied by different competitive strategies, and their receptivity to change is also variable. Implicitly, from the arguments of this and the preceding chapter, organizations which adopt new technology as part of a competitive strategy almost invaiably change their structure: this applies here in particular to industrial relations. Certain propositions follow. If this approach is accurate, one would expect change to be accepted where the necessary conditions of information disclosure, equity, and employee access to decision making exist. One would expect spot contracting to accommodate marginal change but to cause substantial conflict where major changes to processes are proposed. One would expect contingent claims contracts to decay into spot contracts as the pace of change accelerates. However, for technical changes to be important for industrial relations, they must affect some procedural or substantive aspect of the effort bargain. Changes are relevant to the extent that they affect skill levels, employment, work monitoring, health and safety of working conditions, or the relative economic positions of union and management. In some industries, changes have affected all aspects of the effort bargain and its regulation within a very short space of time: the next chapter will focus directly on such cases.

NOTES

1. He assumes that these two forms are essentially individualistic, and contrasts them with the collective bargaining characteristic of internal labour-markets. However, as Fox (1974) has shown, the contrast between individual and collective forms of contract regulation is more apparent than real.
2. The argument against intensive metering runs as follows: 'Efforts to divide the employment relation into parts and assess each separately in strictly calculative, instrumental terms can have, for some individuals at least, counterproductive consequences . . . Rather than regard transactions in strictly *quid pro quo* terms, with each account to be settled separately, they (individuals) look instead for a favourable balance among a related set of transactions'. (Williamson 1975, 256.)
3. This is, in fact, very similar to Gordon's (1976) distinction between qualitative and quantitative efficiency: see Willman (1983a, 126–8).
4. The discussion which follows develops that previously presented in Willman (1982b).
5. Elsewhere he argues that 'failure to respect efficiency arguably reflects power' (1982, 19).
6. This is not a necessary part of the argument, nor need the line be straight or go through the origin.
7. There are also interesting parallels between the characterization of trade unions as a 'voice' mechanism by the Harvard school and Marxist theories about the 'incorporation' of trade unions. However, these parallels will not be pursued here.

6

The Dynamics of Spot Contracting

INTRODUCTION

THE purpose of this chapter is to discuss the ways in which change is negotiated under spot contracting and to look in detail at two of the three cases, docks and national newspapers, where radical process change and a removal of spot contracting went hand in hand. The third case, that of motor vehicles, is different in character and will be the subject of more extended discussions in succeeding chapters.

The chapter focuses on events in docks and national newspapers in both the USA and the UK. The principal intent of this comparison is to show that similar issues arise in the reform of spot contracting in both nations and, thus, that the problems of conflict which arose were not specific to the UK. However, parallels with events in the USA are relevant for two further reasons. In the first place, the USA has a 'pluralist' rather than a 'corporatist' system of industrial relations, and thus shares certain basic features of interest-group organization with the UK. On a more mundane level, change in the two industries occurred first in the USA, and the American example was very much in the minds of negotiators in the UK while changes were bring made.

DOCKS AND NATIONAL NEWSPAPERS

Both dockwork and national newspapers have been the subject of detailed academic investigation. It is thus not the purpose of this section to provide a detailed description of the periods of change in the two industries (although additional material covering subsequent events has been introduced), and readers in search of such a description are referred to the sources noted. Here I shall be concerned to analyse the contractual arrangements which existed prior to change, to discuss the changes themselves, and to show that the strategic motives underlying the change rendered these contractual arrangements obsolete: subsequent contractual arrangements in both industries tended more towards comprehensive contracting. Prior to change, both industries displayed rather pronounced forms of spot contracting which were almost certainly untypical of industry more generally; however, I shall be concerned to show that even more diluted forms of continuous contracting have similar problems.

Dockwork

Historically, work in the ports relied upon a simple system of free-call, in which men were employed by the half-day on specific tasks. Dockers

> would present themselves at the dock gate, or any other recognised hiring point, and foremen would call on as many as the business of the day demanded: those hired were usually kept to the end of each loading or discharging operation before being paid off, and those who did not find work at the main call would wander round the dock in the hope of finding a job elsewhere, or go home until the next day. (Wilson 1972, 17.)

Prior to World War II, the system was characterized by under-employment and poverty which attracted the concern of social reformers (Jackson 1973, 5–22); it also attracted workers who were ready to accept fluctuating earnings and to work for a very large number of employers.[1]

The number of employers reflected economic pressures. In the UK, as elsewhere, dockers tended to work for employers who were agents of shipowners: the port authority which owned the capital equipment did not, under the casual system, engage in stevedoring, primarily because the berth time of the ship represented the greatest unit cost in the unloading process. The pressure thus lay on the ship-owner to secure labour in such a way as to minimize berth time (Jansenn and Schneersen 1982, 25).

Although wartime registration, limiting the number of dock-workers, and the establishment of the National Dock Labour Scheme (1947) went some way towards eradicating social problems, they did not, in fact, eradicate casualism. Under the 1947 scheme, the costs of employment instability were partly transferred from employee to employer. For the majority of dockers, casual employment persisted with the distinction being introduced between the *operational* employer, with whom the docker typically had a task-specific relationship, and the *holding* employer (i.e. the dock labour board), on whom he relied for fall-back payments when work was unavailable[2] (Jackson 1973, 36–7; Wilson 1972, 111–33).

The origins of this system lay in the unpredictability of arrival of ships and seasonal fluctuations in trade combined with the requirement to turn ships around as rapidly as possible. As McCormick puts it,

> Each employer would base his demand for labour on his peak demand and the total demand for labour would be given by the sum of all employers' peak demands. But all peaks rarely coincided, with the result that unemployment occurred. (1977, 256.)

The employers' natural response was to eschew responsibility for labour except when required: the Dock Labour Scheme thus caused a significant addition to labour costs. The dockers' response was classically opportunistic in the face of such indifference: faced with uncertainty not only about the quantity of work but also about its quality—since cargoes could differ substantially in ease of unloading, unpleasantness, and weight—devices were developed to increase the numbers employed, extend the length of work, and maximize wages.

These 'restrictive practices' were of three types:

(i) Those aimed at spreading work across as much of the labour-force as possible, including inflated manning scales, controls on the hiring process, restrictions on working hours, and the continuity rule.[3]

(ii) Those aimed at spreading earnings fairly, such as precedence rules for hatch-work.

(iii) Those which were simply a response to bad supervision, including bad timekeeping, extended tea-breaks, spinning out of work, and 'welting' or 'spelling' which effectively doubled manpower requirements.

These practices caused extension of the time a ship stayed in port and featured as bargaining counters in the continuous day-to-day negotiations over prices and manning for a particular load: breakdown of such negotiations could lead to stoppages which were extremely costly, since they slowed down the turn-around of the ship, and the cost of the entire system was passed on to the merchant or shipowner (Wilson 1972, 212–13; Fadem 1976, 105). Compounded with the frequent failure of employers to provide appropriate equipment and the vagaries of weather, these practices could cause the loss of up to one-third of the working day (Mellish 1973, 77).

Because the ports constituted a bottle-neck in the distribution system, the levels of inefficiency and strike-proneness could be tolerated within limits. However, the system of spot contracting which existed after the war could only survive while it gave sufficient employment and adequate earnings *without* generating costs which port users could not withstand. Throughout the post-war period the industry did cope with recurrent technological change. Improvements to cranes and other dockside equipment occurred together with communications and navigational improvements which made ship arrival more predictable (Knott and Williams 1976, 281–5). As a consequence, although traffic rose in the decade prior to the Devlin report, the register decreased by approximately 14 000 (Durcan *et al.* 1983, 305); the number of dockers unable to find work seldom rose much above 10 per cent in this decade (Jackson 1973, 100).[4]

However the costs of non-productive time continued to rise (Wilson 1972, 250) and the impact of dock strikes on the economy became sufficiently visible for a Committee of Inquiry under Lord Devlin to be set up in 1964 to look at the wide issue of decasualization and efficient working. The Devlin reforms were implemented in two phases, on the theory that previous efforts had failed through trying to do too much at once. Phase I, implemented in 1967, introduced decasualization and secured permanent employment for the vast majority of dockers. However, Phase II, negotiated in 1970, reformed the wages structure and paved the way for technological change.

Although the contractual reforms preceded the major impact of containerization, the two changes cannot be considered in isolation. The cost-reducing pressures which required more efficient labour utilization similarly promoted the adoption of container technology. As Wilson notes,

> it was fundamentally the pressures of transport technology which directed the attention of the Government, employers and trade unions to the reform of labour relations in the 1960s. Imminent containerisation was implicit in the *timing* of the Inquiry . . . (1972, 290.)

He goes on to suggest that Devlin may not have adequately foreseen the full impact of change (1972, 290).

The key problem faced by the shipping industry at the time was the reduction of necessary port time through maximally efficient working to improve the utilization of the most expensive capital investment—the ship (Knott and Williams 1976, 284). Considerable improvements in efficiency could occur through 'intermediate'

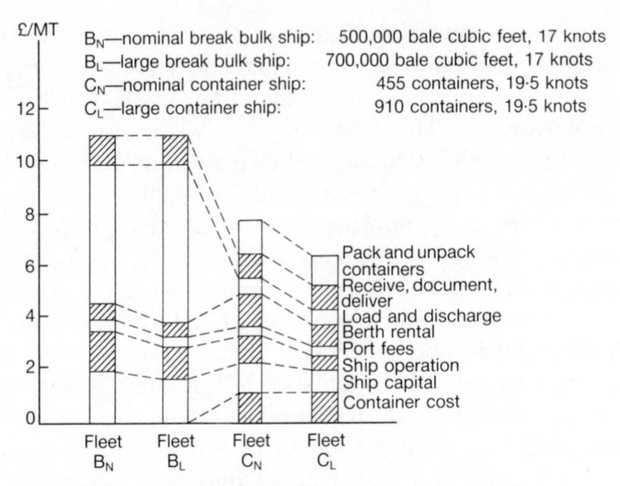

Figure 6.1. Port-to-port-Costs (£ per Metric Ton)
Source: NPC 1967

changes such as palletization[5] or improved labour efficiency but the quantum leap to containerization—where standard containers could be unloaded in 2–3 minute cycles—was considerable. Figure 6.1 shows comparative costings available at the time on a port-to-port basis for different types of ship: these costings have been subjected to some criticisms (Johnson and Garnett 1971, 48–9), but these do not affect the clear reduction in load and discharge time on the introduction of containers.

The prospects for increased labour productivity were also immense. The McKinsey report for the British Transport Docks Board estimated that approximate output per man could be increased from 25 freight tons per man-week on conventional operations to 600 freight tons on containerized cargo, i.e. by a factor of 24. In their example, output increases by this factor to retain approximately stable employment (BTDB 1967, 84–86).

Considerable advantages for *shippers* of commodities also emerged. Faster delivery led to lower inventory costs, allowing higher turnover. Moreover, packaging costs were reduced and pilferage was less likely (Levinson *et al*. 1971, 269). However—and this is crucial for the effort bargain in dockwork—ship-by-ship comparisons deflect attention away from the essential feature. Containerization applies mass-production techniques to goods distribution which requires

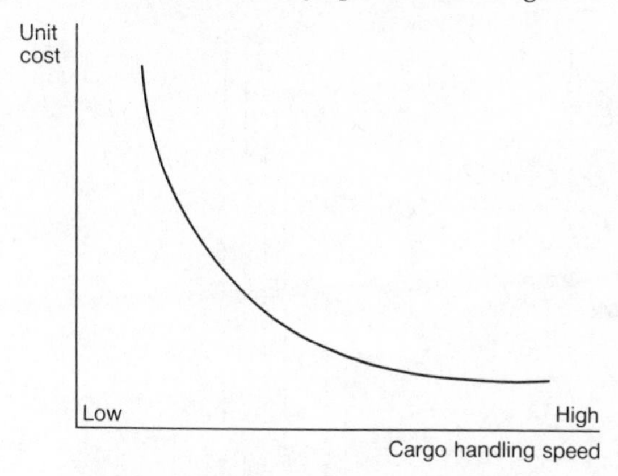

Figure 6.2. Unit Cost and Port Efficiency
Source: Johnson and Garnett 1971

standardization and rapid throughput: capacity utilization thus becomes important. The shipowners' investment in cellular shipping required parallel investment in straddle carriers, side-loading trucks, container cranes, and computerized information systems on the quayside. A high load factor on such equipment (Johnson and Garnett 1971, 31–2), and thus continuity of work, was vital.

Moreover, the gains from improvement of handling speeds were considerable: as Figure 6.2 shows, the cost curve is such that the immediate cost gains to efficiency improvements were substantial. Given the steepness of the curve, the priorities of the Phase II Devlin negotiations were to secure continuity and more rapid turn-around time, rather than to seek manning reductions: as Wilson notes, the immediate cost of labour content was secondary (1972, 250–1).

This is important for the period following decasualization. Given that Devlin Phase I merely established permanence without reforming the spot-contracting system for prices and manning, one might expect that opportunism would persist. In fact, wage drift over the 1967–9 period continued at a high rate. Mellish shows for London docks both that earnings of London dockers almost doubled during the 1967–9 period and that employers were able, in the short term and in the knowledge that Phase II was imminent, to pass on these increases as increased charges to customers (1973, 23, 42). The issue was to purchase a system which would secure continuity rather than to exert tight control over labour costs. Moreover, as Figure 6.3

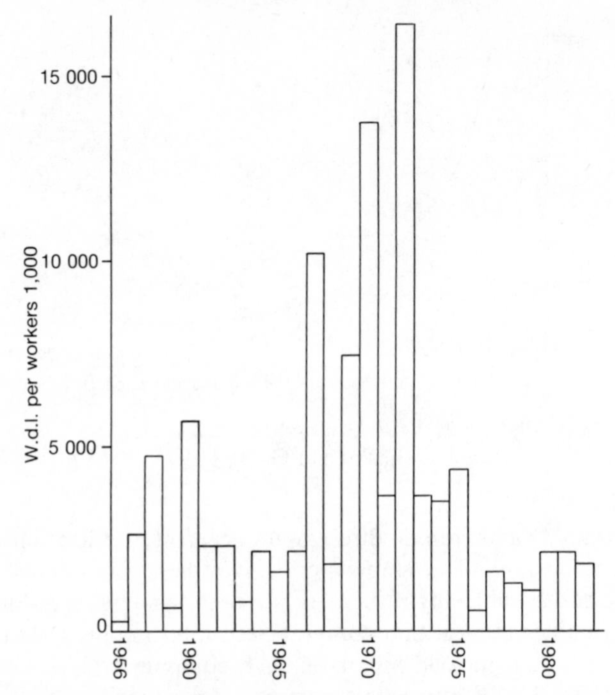

Figure 6.3. Working Days Lost per 1000 Workers in Port and Inland Water Transport, UK 1956–1982
Source: Durcan *et al.* 1983; DE *Gazette*

shows, strike activity increased markedly during the period: in 1967 a wave of strikes took place over the price of decasualization, and in 1968 and 1969 a number of wage disputes occurred over piece-work prices from the stronger platform of permanent employment. As Wilson puts it,

> the difficulty of separating reform of the system of employment from reform of the wages structure was that *both* contributed inextricably to the inefficiencies of dockland. Wasteful practices were the response to the insecurity of casual hiring and to insecurity of earnings under piecework: and the removal of one form of insecurity without removal of the other was bound to lead to some form of pressure from the shopfloor. (1972, 51–102.)

During this phase, a ban was imposed upon the introduction of new technology until Phase II was negotiated: agreement on the latter was forthcoming only after a national dock strike in 1970 led to an increase in the price of change.

The Phase II agreements themselves are of some interest. In London, the agreement was contained in a 'Green Book'. Its stated aims were to promote efficiency, facilitate mechanization, and improve terms and conditions of employment: it required dockers to observe certain work practices, including mobility and flexibility, and to achieve a satisfactory level of output, so remedying the tendency towards effort restriction under casualism. Employers had the right to place men in particular jobs, and the payments for such jobs were set according to a scale of 'contingency' payments.[6] Shift working was introduced, a system of formal penalties for lateness was established, and manning arrangements were set (Mellish 1973, 51–102).

All of this is, of course, a comprehensive contract suited to capital-intensive standardized processes. Mechanization destroyed the 'discreteness' of tasks, since it allowed loading, unloading, and storage to be combined. It permitted individual, as opposed to gang, working, and it required increased discipline, co-operation, and flexibility of labour without the possibility of disruption through disputes over earnings variation.[7] Some variance continued between ports: Liverpool, among others, retained some piece-work (McCormick 1977, 257) while elsewhere the Phase II agreements merely codified existing practice (Fadem 1976, 29).

However, the comprehensive contract did not spell the end of conflict over change. As practically all commentators on the period note, the success of reform merely accelerated the movement of work away from areas controlled by registered dockworkers—this time on a cost basis. When work moves inland, dockworkers find their 'automatic entitlement to work has become a virtual disqualification' (Wilson 1972, 195) because of their higher wages. Container operators have strong incentives to move the still labour-intensive process of stuffing and unloading containers inland, just as dockers try to 'chase' such work and have it registered as dockwork. Container-ban strikes occurred in 1969 in Liverpool and London, were central to the national dock strike in 1972, and reappeared in London in 1975.[8] Long-term blacking of containers occurred: for example, generally between 1968 and 1970, and at Hull between 1974 and 1976.

Devlin had not explicitly considered the definition of dockwork, but subsequently the issue was forced high on the agenda: the 1969 Bristow Committee suggested a five-mile corridor on either side of the Thames within which all container work was to be done by registered dockers, and later the abortive Dockwork Regulation Bill attempted to delineate a similar 'cargo handling zone' around all ports. Both took USA experience as their model, but neither was successfully implemented.

Since containerization threatened to 'eliminate the dockworkers from the chain of transport' (Fadem 1976, 313), contractual reforms were jeopardized from the outset. Opportunism and conflict were unlikely to disappear since behaviour 'rooted in the insecurities of casualism could flourish just as easily in the insecurities of declining job opportunity' (Wilson 1972, 304). Employers had guaranteed 'no redundancy' prior to the Devlin reforms and had successfully operated a system of voluntary severance (Wilson 1972, 249, 264; Durcan *et al.* 1983, 295). However, the rapidity of the fall in employment (see Table 6.1) has put strains on severance schemes: they have been periodically improved in the 1980s by the input of public money, but the continued employment of dockworkers has caused some disputes, notably at Liverpool in 1980.

A more detailed assessment of the implications and successes of reform can be gained from research publicized by the National Ports Council between 1971 and 1981. The most general point emerging from the regular 'bulletins' is the massive growth in container traffic, particularly between 1970 and 1974, when tonnage volume almost doubled. This was reflected in movements in numbers employed: within the area covered by the National Dock Labour Scheme, large ports with traditional concentration on break-bulk general cargo suffered substantial employment reductions: for example, between 1965 and 1975, 61 per cent at London, 46 per cent at Liverpool, and 71 per cent on the Clyde (NPC *Bulletin*, No. 9, 1976, 37, 51–5). However, Southampton and the East Anglian ports fared much better: by 1979, growth in the number of non-registered dockworkers involved in work associated with North Sea oil and roll-on/roll-off services to the Continent was contrasted with accelerated contraction in the registered force in an employment census (Bulletin, No. 16, 1981, 17.) Dissatisfaction with the relatively easier environment and higher status of container work amongst those who continued to work in break-bulk areas persisted (Fadem 1976, 54).

One crucial area of employment outside the registration scheme was in plant maintenance: 1976 figures revealed that demand for fitters in containerized ports was 85 per cent up on conventional requirements (Bulletin, No. 8, 1976, 10). High labour turnover, insufficient training, and the retention of demarcation between trades all contributed to difficulties in dealing with shore equipment utilization problems, particularly on straddle carriers where down time in the mid-1970s could approach 50 per cent. (Bulletin, No. 8, 1976, 5–34; No. 14, 1980, 46–69). The introduction of new equipment posed several problems for the retraining of foremen, and for control over mechanical-equipment operators who apparently retained traditional preferences for 'welting' systems (Bulletin, No. 4, 1973, 33; No. 14, 1980, 46–68).

In fact, restrictive practices have survived in several ports and have been transferred to new equipment. While some ports by 1979 had substantial flexibility in a permanent work-force, others had written agreements restricting deployment of labour: overmanning, refusals to change functions on a shift, lack of gang mobility, restrictions by some drivers to one-job working, demarcation between trades, and paired maintenance working were most damaging. As a consequence, while variance in containers handled per item of equipment was low in 1979, in a sample of six ports with restrictive practices, output per man was between 127 and 374 containers per man-year; where no such practices existed, output of 782 per man-year was achieved (Bulletin, No. 16, 1981, 1–15).

In part as a consequence, UK performance lagged behind that of competitors. In 1978, the average berth time of a sample of UK ports (several of which were outside the Dock Labour Scheme) was 44 hours, compared with an average of 26 hours for a Continental sample. Rotterdam, a major competitor, was achieving turn-around in 22 hours (Bulletin, No. 15, 1980, 22). The UK times were undoubtedly lower than those of a decade earlier, but the competition had not stood still.

Table 6.1 shows the trends of output and employment for UK ports from 1965 to 1980. The huge productivity improvement implied by the two trends is more apparent than real given the growth of unregistered employment. Nevertheless, the protracted and sharp fall in registered employment is clear: although the largest single falls followed full decasualization and the container strike, a steady loss of employment characterized the late 1970s.[9]

In terms of the approach outlined in the previous chapter, work on the docks before Devlin was essentially a sequential form of spot contracting conducted under conditions of considerable uncer-

Table 6.1. Output and Registered Employment in Dockwork 1965–1980

Year	Traffic[a] (Index 1965 = 100)	Registered work-force[b]
1965	100	65 522
1966	103	62 522
1967	106	59 383
1968	115	56 563
1969	121	52 732
1970	129	46 912
1971	133	45 491
1972	135	41 247
1973	147	31 662
1974	139	32 423
1975	119	32 243
1976	128	30 144
1977	125	28 746
1978	129	28 708
1979	140	26 573
1980	132	23 010

Sources:
[a] NPC *Annual Digests* 1966–79; *Port Statistics* 1980.
[b] NDLB *Census*, annually.

tainty. Employers were characterized by indifference towards casual employees, and employee opportunism in the form of restrictive practices backed up by the threat of strike was rife. Although the system could cope with day-to-day change where ship turn-around was more important than labour cost, in the long term the delays promoted by the contractual system stimulated its demise. In the short term, spot contracting was a suitable response to product-market variance; in the longer term, the needs of shipowners for rapid job finish ran into conflict with the labour-market necessities of job expansion and creation. Technological change was clearly of the cost-minimizing variety, and prompted the adaptation of an effort bargain suitable for mass-production techniques. A contingent claims contract was established, to be jeopardized by the lack of appropriate foresight on the part of port employers. They failed to anticipate the extent of reduced labour requirement and the consequently increased costs of severance; restrictive practice continue.

A number of further points are of interest here. Gang organization, the absence of supervision, and the nature of casualism combined to produce a very high degree of lateral, and low degree of vertical, trust. Consequently, although decasualization offered security and higher earnings, dockers tended to cling to opportunistic forms: the *atmospheric* consequences of casualism jeopardized the move to a stable contractual form or, as Wilson puts it, 'the practices of casualism outlived the casual system itself' (1972, 190). During the change, the emphasis of trade union activity was on job control rather than on remedial efficiency-improving activity. Collective bargaining could not act as the vehicle for change without conflict: the docks were characterized both by inter-union strife and by the activities of unofficial organizations clearly beyond the control of the formal trade unions (Wilson 1972, 192–212; Jackson 1973, 47–57; Durcan *et al.* 1983, 284–6). While divisions between and within unions formed no stable and clear-cut lines, it seems fair to say that commitment to the reform of casualism was typical of full-time officials of the TGWU while the unofficial movements, and to some extent the rival union, the NASD, tended to press for advantages *within* the overall framework of casualism.

The third point is that dock employers tended to be pushed into change. The stimulus to containerization came from elsewhere, and although the example of the USA (see below) was there to draw upon, port employers tended to follow rather than lead; since they faced considerable uncertainties themselves as a consequence of the Devlin reforms, this general lack of foresight did not assist the development of trust on the part of employees.

I shall return to these points, and to the development of comparisons with the USA, later in this chapter. However, the next section focuses on an industry which experienced similar problems in a very similar institutional context.

National newspapers

The system of employment in national newspapers bears a number of similarities to the system operating in dockwork, particularly that of the period before the Devlin reforms. The trade unions are the sole supply of labour, i.e. there is a labour-supply closed shop operated by the several unions which organize the area.[10] In addition, the labourforce divides into 'regular' and 'casual' employees, albeit in rather different proportions from that in pre-Devlin dockwork. In 1975, 10.4 per cent of employees in London and Manchester were casual, primarily employed in the machinery and publishing departments where labour demand was particularly variable: this is apparently a

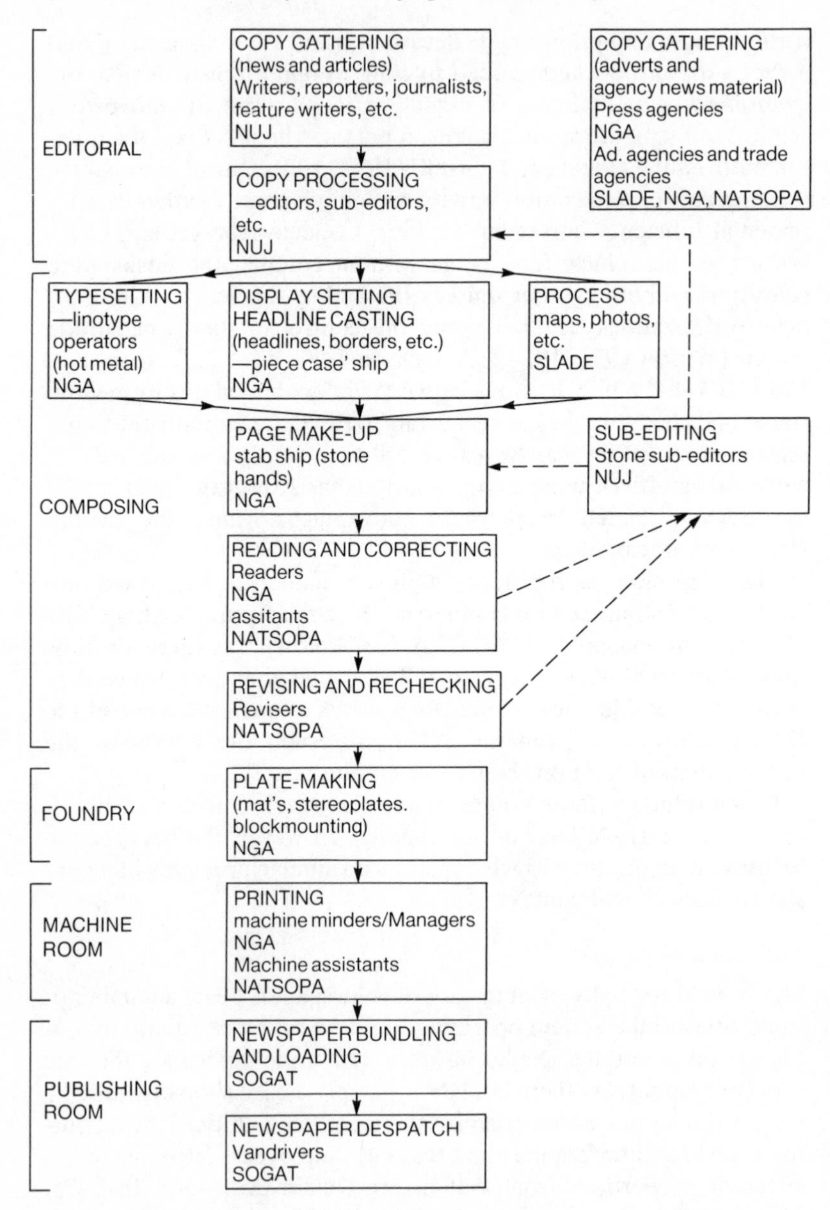

Figure 6.4. The Organization of Production in a National Newspaper
Source: OU 1976, 14

rather smaller percentage than in the past, owing to the signing of 'decasualization agreements' (Royal Commission on the Press 1976, 56; Sisson 1975, 77).[11] A second difference from the dock labour-market was the existence of strict union demarcation lines between jobs in the production process. A simplified flow diagram of the organization of production appears as Figure 6.4 where it is clear that, *within* stages of the process, responsibility for tasks is divided between unions; moreover, distinctions between different chapels *within* unions complicate the situation further.

Although employment insecurity may be restricted to casual workers, there is substantial insecurity of earnings on national newspapers. This follows from the prevalence of 'payment by task' for both regular and casual workers: industry basic rates tend to be a small proportion of total earnings, the greater part being made up by the addition of extras paid only when particular tasks are performed, and even this small proportion showed substantial overall decline in the period 1969–75 for most occupations (ACAS 1976, 256; see also Sisson 1975, 43).

The development of this system of employment and payment is related to the peculiar nature of the product. As Reddaway first noted, newspaper proprietors actually operate in two related product-markets: advertising space is sold to advertisers on the basis of the sale of copies to readers. The demand for the former is related to copy sales, but the differentiation of newspaper readership itself helps to sustain the distinction between 'popular' and 'quality' papers: 'popular' papers have high circulation levels, relatively low price, and relatively low advertising revenue per reader, whereas 'quality' papers have lower circulation but are able to charge higher advertising rates—and thus gain higher advertising yields per unit of circulation—because of their access to potential consumers with higher disposable income (Reddaway 1963, 201; Royal Commission on the Press 1976, 23–5).

However, only one product is produced in both categories, and it is highly perishable. Moreover, the product-market is subject to several different sources of variation:

(i) Long term: dependent upon the overall level of economic activity and competition between newspapers and other media; the former influences the overall level of advertising expenditure and the latter the proportion of it which goes to national newspapers.

(ii) Medium term: resulting largely from the success of editorial policy.
(iii) Seasonal: stemming from peaks and troughs in consumer expenditure.
(iv) Weekly: advertisers favouring the period between Wednesday and Friday.
(v) Daily: depending upon the type of events which require reporting. (OU 1976, 47–9; ACAS 1976, 27.)

The existence of this uncertainty gives a payment-by-task system considerable advantage in the context of a labour-supply closed shop. As Sisson notes,

> Because of the many changes in the product and fluctuations in the level of activity in the product markets, the overriding need is to keep *unit* production wages to a minimum. Since the managements are unable to adjust to changes in the demand for labour in the usual way, as system of payment by task has one obvious advantage: it ensures that payment is only paid when the task is performed. (1975, 149.)[12]

It also has certain competitive consequences. Despite substantial differences in the pay of occupations between offices, the system of localized task-based bargaining helps to take unit production wages out of competition. The system of casual labour reinforces this, while union control over labour supply makes entry into the product-market more difficult (Sisson 1975, 157, 170).

One of the more comprehensive systems of task pricing is used to cover the work of linotype operators in the composing room. The London Scale of Prices, based on national agreement, sets out minimum payment for lines of various length of various sorts of type. Although this leads to calculation of earnings based on individual performance, payment is in fact based on a pooled average. Group earnings thus depend on a reasonably rapid pace of work and maintenance of fairly low manning levels. In form, then, the scale is a contingent claims contract, similar to the system of contingency payments on the docks. However, as with the dockwork case, the scale is in fact subjected to recurrent bargaining: 'when there is a need for a new job to be priced, the LSP often only provides general guidance' (ACAS 1976, 49).

This is, in fact, typical of wage settlement throughout the industry. National newspapers engage in daily product change, effected at very short notice, and a contingent claims contract cannot specify all eventualities, particularly in the face of a definite interest on the part of local union organizations in attenuating the effects of product variation on employment and earnings.

> It is the discontinuities in production which pose the biggest day-to-day threat to the interests of chapels . . . All these discontinuities disrupt working arrangements. Before the negotiation of comprehensive agreements, they threatened the stability of earnings as well . . .
>
> This type of situation makes it essential for the chapels to seek some control over the impact of changes in the product and fluctuations in the level of activity in the product market. But it also gives them the ability to do this simply and effectively: all they have to do is insist that the managements negotiate with their representatives whenever these changes and fluctuations result in a need to alter working arrangements. (Sisson 1975, 102.)

This insistence occurs frequently with the result than an *ad hoc* system of sequential spot bargaining has been the norm (Martin 1981, 47).

The system has a number of expected characteristics. The first is that, as with dockwork, local union organizations were relatively successful in pursuit of policies of earnings stability and employment maximization, both of which were essentially in conflict with product-market characteristics. In order to achieve such goals, a basic principle was the establishment of mutuality—no change without prior negotiation and agreement—in relations with all printing houses. The second was the development of a number of 'restrictive practices' these included a 'blow' system analogous to the 'welt' on the docks, overmanning, restrictions on output, and payment for overtime not worked, as well as the many restrictions on flexibility and mobility of labour inherent in demarcation arrangements (ACAS 1976, 58–9, 184, 296; Martin 1981, 53, 240; Sisson 1975, 112–43). Thirdly, and as a consequence, the form of labour organization practised up to the mid-1970s was essentially a modified inside-contracting relationship between managers and chapels: the latter were responsible for recruitment from the relevant local branch, manning arrangements, holiday, shift, and overtime rotas, certain promotional decisions, and disciplinary matters. In addition, all relationships between employer and employee, save purely personal issues, had to be dealt with by chapel officials (ACAS 1976, 129–33).

As with dockwork, this essentially inefficient system required particular conditions to survive. Many national newspapers in the early

1970s were owned by groups with diverse interests which did not require profitability from institutions which attracted substantial prestige, or by interests which for other reasons were prepared to allow for losses as the cost of ownership (Royal Commission on the Press 1977, 20–9).[13] However, a crisis emerged in the early 1970s owing to long-term difficulties in both readership and advertising markets. These were exacerbated both by the sharp cyclical downturn affecting the UK economy after 1973 and by a rapid rise in the price of raw materials, particularly newsprint (Royal Commission on the Press 1976, 23–31; Martin 1981, 1–15). Newspaper producers reacted to product-market decline by more intensive marketing and by continuing with a strategy of diversification into other media. However, one of the major areas of concern was the reduction of costs. As Martin notes, this inevitably led to a focus on labour costs:

> Attempts to reduce costs involved attempting to increase efficiency in the use of raw materials, especially newsprint, and, more importantly, to reduce labour costs, mainly through adopting new technology. (1981, 15.)

Since raw-material utilization was efficient anyway, and newsprint costs were outside company control, only one option remained.[14] Moreover, as Martin notes, given the decentralized bargaining structure the major scope for competitive advantage in national newspapers lay in the reduction of labour costs in-house (1981, 205).

The cost structures of newspapers were examined in a number of ways by the Royal Commission on the Press in its publications. In the most general comparison, production wages were a stable element in cost changes between 1970 and 1975 at 24 per cent of the total; the most substantial increase during the period was in the cost of newsprint, which increased from 26.5 to 31.3 per cent of the total. Overall, wage and salary costs were stable as a proportion of the total over the period at about 43 per cent (ACAS 1976, 266); the global labour-cost figure comprises the largest single element in costs. Moreover, it is likely that composing-room wages made up a substantial proportion of the total production-wages cost: Table 6.2 presents a calculation of likely average wage costs for 1975 as a share of the total wage bill by department. It reveals that the main stages in the production process incurred approximately equal wage costs. However, other things were not equal, and it was always likely that attention would focus on the composing room.

One factor at work was the history of investment activity. A major breakthrough in printing technology in the 1950s had been the

Table 6.2. Wage Costs in National Newspapers, 1975

Type of work	Total no. employed[a]	Percentage of wage bill[b]
Composing and Related	4654	28.5
Machine room	6221	28.6
Publishing room	6686	23.8
Maintenance	2285	10.7
Other	3180	8.5

[a] Regular workers in London.
[b] Based simply on a multiplication of employment figures in column 2 by the earnings average for the various areas.
Source: Royal Commission on the Press 1976, 56; ACAS 1976, 249.

development of web-offset printing methods, giving improved quality of reproduction; however, in the 1960s Fleet Street invested heavily in letterpress machines, the replacement of which would have been uneconomic. Moreover, although manning levels could doubtless have been revised downwards with technological change in the machine room, web-offset techniques were extremely expensive, would have required the closing-down of production during introduction, and, in any event, tended to improve quality rather than reduce costs (Martin 1981, 28). Developments in 'paper-strangling' (i.e. automatic binding) and mechanical handling were gradual rather than radical, and offered no substantial economies in publishing areas. Attention thus focused on the composing room.

The basic technology of the composing room had not changed substantially for approximately a hundred years. The essential operation consisted of the production of lines of typeset material using molten lead to form a relief of the material to be printed (Martin 1981, 26–31; Sisson 1975, 8–12). The operation was labour intensive, reasonably flexible to late changes in copy, but expensive, owing to the earnings of linotype operators. Alternative means of composing with the use of computers had been in existence since the 1950s: for example, on the *Guardian*, NGA operators stroke keys to produce punched paper tape (rather than lead) which is then fed into a computer for hyphenation and justification, and subsequently into a hot-metal process. However, the more basic change was the emergence of photocomposition in the 1970s.

The basis of photocomposition has been described in several publications.[15] Martin's description is as follows:

The new technology of computerised photo-composition as it developed in the 1970s provided the possibility of ultimately merging editorial and composing functions, with large reductions in composing room staffs and the simplification of the production process. In the most advanced systems, copy can be keyed directly into the computer, bypassing the compositor, the computer can be programmed to permit page make up through visual display terminals and whole pages can be photocomposed: in such systems, plates can be made from camera photographs of page bromides and passed to the machine room for printing. (1981, 30.)

Case histories of provincial press applications of the technology by the Royal Commission revealed the possibility of 30–40 per cent reductions in composing costs. National newspaper publishers felt at the time that composing-room employment could be reduced by 50 per cent. American examples showed that the cost of investment had been recouped within two years. In the UK, the Royal Commission estimated that an investment of £20 million in such techniques would save £10 million a year in labour costs (at 1975 prices) (Royal commission on the Press 1976, 53–4). Photocomposition was thus a key remedy for low labour productivity[16] and was more important for quality papers where typesetting costs were higher than for popular ones (Royal Commission on the Press 1976, 25).

Just as containerization could be seen as a way to 'bypass' the dockworker altogether, photocomposition could be used to 'bypass' compositors. At least two ways of using the system could be considered. The most radical involved the origination of newspaper text by journalists—variously described as 'front-ending', 'single-keying', or 'direct inputting'. The NGA, organizing compositors, decisively rejected such systems but did *not* reject in its policy development the use of new technology within existing demarcation lines where 'the input of typographical material is a composing function' to be performed only by union members.[17] Basically this policy hinged around maintenance of control and of job security (Martin 1981, 74–114).

However, reform of the employment relationship was essential. As with dockwork, the spot-contracting system had been able to accommodate piece-meal change within the framework of a constant overall production system: mechanization of activity in publishing departments, teletypesetting in the wire room, and recurrent improvements to inking and machine speeds in the machine department had all been negotiated via mutuality—often on the basis of employment reductions and the sharing of wages (ACAS 1976, 39; Royal Commission on the Press 1976, 46; Sisson 1975, 60, 110–3, 128–9). However, the LSP could not withstand radical change:

Existing wage systems would be inappropriate for new integrated systems because output would be dependent upon the speed and output of the technology not the efforts of the individual operative, and it would be impossible to link effort to output in systems where input could be coming from three or four different sources. (Martin 1981, 96.)

Again, as with dockwork, the attempt to move away from spot contracting preceded radical innovation. Sisson notes the development of so-called 'comprehensive agreements' from the 1950s. Often associated with technological changes such as mechanization in publishing areas, they tended to codify practices, introduce cost controls, and, occasionally, provide a single wage for a 'comprehensive service' (1975, 143–5). However, Martin notes that several such agreements suffered from the expected defects of imperfect contingent claims contracts: they failed to specify coverage of items and tasks, leaving loose ends to be bargained over; and they encouraged restrictive definitions of work assignments in the expectation of such bargaining (1981, 49). However, the same cost-control considerations which encouraged photocomposition encouraged prior tinkering with contractual arrangements.

The first period of change has been discussed by Martin (1981) in some considerable detail; only the salient parts are relevant here.[18] As with dockwork, initial industry-wide attempts to accommodate radical innovation foundered on the resistance of employees and local union organizations concerned to retain spot contracting. A comprehensive set of proposals intended to reduce employment and earnings insecurity in the industry emerged from the deliberations of the Joint Standing Committee for National Newspapers in December 1976. Published as *Programme for Action*, it contained proposals on redundancy, decasualization, and pensions, together with a planned industry-wide disputes procedure and a programme for the acceptance of new technology. It was decisively rejected by Fleet Street members of all unions early in 1977: the interests of chapels on Fleet Street, of national unions which also represented employees with aspirations to work in national newspapers, and of employers concerned with cost control could not be reconciled.

Subsequently, house-level deals have been sought in a number of newspapers. At Mirror Group Newspapers, the company bought out the LSP and established a single composing-room rate. The deal provoked resentment over the disturbance of traditional differentials among journalists and machine-room staff, resulting in disputes in

1978. More disturbingly, major technological difficulties emerged in the transfer of publications to computerized photocomposition: for some employees, the higher rates of pay on the LSP were retained. Technological change did not lead to the creation of a more unified and flexible labour-force, nor to a reduction of labour costs in the short term.

Nevertheless, the agreement was more comprehensive than previous examples of the genre. It allowed for a reduction in composing-room staff from 267 to 170, the latter number being stable up to a 5 per cent increase on agreed pagination, and confirmed that replacement of staff would not be automatic. However, it looked rather warily towards a future characterized by continuous negotiations. The agreement was set to last for four years, but it included the following provision:

> During the lifetime of the agreement, there may both minor and major changes in technology. In the event of major change of technology or substantial change or production requirement of any or all titles, the management and NGA negotiating panel shall meet to negotiate the changed circumstances.[19]

At the *Financial Times*, radical plans in 1975 for direct inputting gave way in the face of both technical and social problems to a piece-meal approach, subsequently to no action at all, and to postponement of innovation in 1977. Subsequent attempts at cost cutting have tended to focus on manning levels in the machine room, the issue which prompted the ten-week strike in 1983.[20]

However, perhaps the major focus of interest has been on events at *The Times*, which experienced a year-long dispute (from November 1978 to November 1979) over the introduction of new technology. The plan for change originated in 1976, and from the outset the goal was direct inputting on all Times Newspapers Ltd. publications as well as substantial manning economies in downstream operations in machining and publishing houses. Once more, a UK newspaper experienced substantial embarrassment through the adoptation of systems appropriate to newspapers in the USA but ill suited to their own. In April 1978, *Times* management issued an ultimatum requiring introduction of new technology, new manning levels, wage restructuring, the ending of restrictions on production and of unofficial action, and a new disputes procedure. In November 1978, after a period of recurrent industrial action, suspension occurred.

The agreement resolving the dispute was in many ways unsatisfactory. A single composing-room rate was established, and flexible working practices promised; 40 per cent manning reductions were agreed for the composing room, and the new equipment was to become fully operational within two years. However, the details both of the operation of new technology and of manning reductions in the machine room were vague. As a result, production of Times Newspapers Ltd. publications by new methods did not occur until after ownership of the paper had changed hands. Under the ownership of Rupert Murdoch, an agreement in March 1982 secured 600 voluntary redundances and the abolition of 900 casual shifts, once more under threat of closure. In December, the issue reappeared with a dispute by maintenance staff over the rate for new electronic work. As Routledge notes, the 1978–9 dispute had a number of interesting consequences. As well as acting as a signal of the difficulties of achieving change, it consolidated union strength in the old practices in a number of ways. First, wages had increased in rival titles as the scramble to win *Times* and *Sunday Times* readership induced higher circulation and pagination. Secondly, two other titles, the *Daily Express* and the *Observer*, both concluded agreements on new technology which accepted the NGA monopoly on key-stroking (Routledge 1979).

Overall, therefore, the implementation of new technology in national newspapers has been substantially delayed by union resistance, in the form both of strike action and of the imposition of costs (severance payments and retirement agreements). It is of some interest that publishers appear to have sought economies in the machine room rather than direct confrontation with compositors.[21] Subsequently, events elsewhere in the printing industry outside the ambit of the Fleet Street labour-market have proceeded apace: in general printing, the NGA has reached agreements on the reform of apprenticeship arrangements to accommodate technological change and on the use of word processors. In provincial newspapers, a major concession on the part of the NGA has been the concession of the principle of single key-stroking in return for guarantees of no redundancy, some common action between the NGA, SOGAT, and the NUJ, elimination of traditional demarcations, and the maintenance of a closed shop.[22] Within printing overall, as Figure 6.5 shows, the *Times* dispute and the BPIF dispute over the shorter working week remain unusually high peaks of conflict.

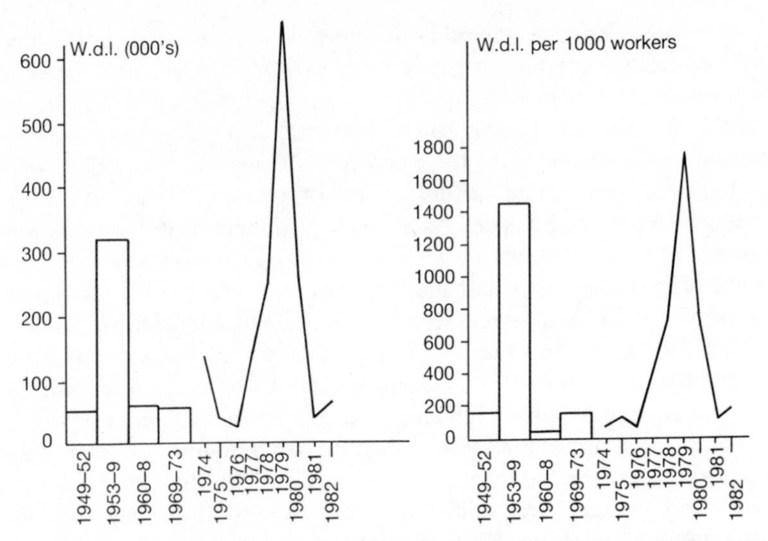

Figure 6.5. Working Days Lost in Disputes in Printing and Publishing, UK 1949–1982.
Sources: Durcan *et al.* 1983; DE *Gazette*

In their discussion of this industry, the Open University Course Team are concerned to emphasize the geographical concentration, the relatively small number of employers, and the presence of a core craft skill in their explanation of the differences between national newspapers and dockwork: the former industry has, they argue, been characterized by much greater employee control, without the additional support of legislation to curtail the number of available employees (OU 1976, 55). However, from the perspective of the previous chapter, the similarities are more important than the differences. In both industries the contractual progression is from a sequential spot contract, where the small-numbers condition on the employee side does not exist, to a contingent claims contract, where the employee suffers the effects of product-market variability. Subsequently, bargaining pressure reasserts a form of continuous inside-contracting which eventually succumbs (or looks likely to succumb) to cost-minimizing technological change. In both industries, the collective bargaining structure is insufficiently robust to cope with the negotiation of change, since employee-controlled forms of continuous contracting emphasize the difference between sectional and general interest. In both industries, this failure contributes to the manifestation of government concern in the form of committees of inquiry or legislation. In national newspapers, as in dockwork, the contours of trust are predominantly lateral: compositors trust each other's performance to the extent of pooling earnings, but are suspicious of all managerial attempts to improve security of earnings. As Sykes (1967) noted, the pattern of union organization overlies patterns of work-group solidarity. Neither pattern of technical change and contractual reform has substantially altered trust relations in either industry.

Both changes bear some of the hallmarks of cost-minimizing change. Cost-cutting exercises are also high on the agenda in areas unaffected by technological change. The innovation itself comes from outside the industry and is imperfectly tailored to it. Managers with no expertise in innovation make sub-optimal choices which affect the level of employee confidence. The focus on labour reduction rather than on output expansion reinforces poor trust relationships, and opportunistic practices are retained on new equipment: this does not mean that the changes are not worthwhile, but some potential gains are forgone.

It may always be argued that such generalizations subsume idiosyncratic features of particular industries. However, I would argue that these two industries do illustrate important aspects of contractual change or 'trust dynamics'. Sequential spot contracts accommodate small-scale change well, but radical innovation which causes work to become machine paced requires a totally different contractual form. Still, it may be argued that the labour- and product-market conditions of such industries are so untypical that it is unduly risky to base generalizations upon them: the following sections begin to assess the general applicability of the argument so far.

GENERAL APPLICABILITY

The first, and perhaps most obvious, point to make is that the pattern displayed in the UK newspaper and dock industries is not untypical of events elsewhere. In particular, both industries could reflect upon the prior experience of similar changes in the USA.

Longshore work

Labour-market organization in the US longshore industry displays the same basic features as its UK counterpart: historically, casual work 'with work guarantees limited to a portion of a day' has been associated with widespread employment and earnings insecurity, the development of work-sharing and 'make-work' techniques on the part of the labour-force, the attempt of labour unions to restrict access to longshore work, and a reluctance on the part of the employers to make investments in capital equipment (Levinson *et al.* 1971, 274–6; Hartman 1969, 25–72). The post-war period on both West and East Coasts has been characterized by increasing use of income and employment guarantees in return for flexibility of working and acceptance of containerization. However, the trajectory of change on the two coasts, which have historically retained separate bargaining arrangements, and the extent of conflict reflect substantial differences in contractual arrangements and the extent of employment security.

On the West Coast, measures limiting the extent of casual working have a long history. Until 1934, recruitment to the industry was via employer-run hiring halls in each port; after an 83-day strike in 1934, the International Longshoremen's Association, soon to separate from the East Coast as the International Longshoremen's and Warehousemen's Union, established coast-wide bargaining and joint control of access to longshore work through jointly administered hiring halls concerned to equalize earnings and employment opportunities. However, the employers remained free 'without interference or restraint from the International Longshoremen's Association to introduce labor-saving devices' (Levinson *et al.* 1971, 285–6).

In the period after this commitment was established, it was subjected to continuous local assault in the form of restrictions on labour-saving techniques or organization applied by longshoremen in defiance of the agreement. However, the post-war period was not characterized by the accelerated development of such practices. Following a strike in 1948, a period of co-operative relations between employers and unions, involving three-year agreements and consultation over major changes, culminated in discussions from 1957 onwards on the possibility of 'buying out' restrictive work rules preparatory to the introduction of new technology: Goldberg sees this as the consequence of 'conscious' and 'overt' decision making by the employers' confederation—the Pacific Maritime Association—to establish a long-term programme for the industry.[23]

The product of such discussion was, in turn, the well-reported Mechanization and Modernization (M&M) Agreement of 1959

(Killingsworth 1962; Hartman 1969; Goldberg 1973). This provided for the 'sale' of all work rules except those concerned with safety in return for a system of job and earnings security over the seven-year period 1959–66.[24] In a system which pre-dated, and perhaps influenced, subsequent UK practice, inducements for early retirement and a guaranteed 35-hour week were offered to employees, together with guarantees of ILWU jurisdiction over dockwork and a sharing of the benefits of change. The agreement was successfully implemented, and renegotiated in 1966: the major problems of containerization were to re-emerge in the strike of 1971 which encompassed both West and East Coasts.

However, the history of change in the East was very different, particularly in the post-war period. Whereas ILWU members were dispatched from the hiring hall, the East Coast, particularly the largest port of New York, operated the so-called 'shape-up' system were

> regular workers present themselves at piers to be checked in and casual workers present themselves at the appropriate employment and information centre, or at the pier, to be selected. (Goldberg 1973, 254.)

This system was very similar to the UK 'free-call' system prior to decasualization in the post-ward period; it was sustained in an environment of employment instability and rank-and-file revolt against the policy of the recognized union—the ILA. No coast-wide bargaining structure was established, and the New York Shipping Association (NYSA) remained the pace-making employers' association for the rest of the Coast (Levinson *et al*. 1971, 277–85). The East Coast was also characterized by government concern and recurrent strike activity. In 1953, following concern with criminal activity on the waterfront, the Bi-State Waterfront Commission was established: it quickly established compulsory registration of longshoremen against the wishes of the ILA. However, this did not signal the end of waterfront conflict: there were major strikes in 1956, 1959, 1962, 1964, 1966, and 1968; five of the stoppages between 1956 and 1968 stimulated the establishment of Taft–Hartley Boards of Inquiry (Jensen 1974, 95–120, 477).

Discussions over technological change reflected this rather different background:[25] the NYSA and the ILA concluded an automation agreement allowing the use of containers provided they were handled by ILA labour. However, it made no provision for an 'M&M' style abolition of work rules, the issue which occasioned the 1962 stoppage on contract expiration; part of the resolution to this dispute was a study by a Department of Labor team of manpower utilization, job security, and related problems. The team discovered substantial inefficiencies in the utilization of labour, but suggested that change could occur on the basis of employment guarantees through reliance on entry limitation and early retirement to stabilize the size of the workforce.

This third-party intervention did not end conflict over change. Disputes in 1964 and 1968 were settled by agreements which reinforced the gradual approach to change: the 1964 settlement guaranteed 1600 hours pay per year in New York, together with better pension arrangements, in return for improved flexibility and a reduction in gang size; the 1968 settlement raised the New York guarantee to 2080 hours, offered a uniform wage rate, and included a clause covering container work for the ILA. One of the enduring contributions to this pattern of conflict was the seniority system in New York.[26] The reforms of the 'shape-up' established by the Waterfront Commission had resulted in a pier-based seniority system which discouraged movement of labour: priority was given by seniority to workers at a given pier but, if they were forced to move away from that pier, industry-wide seniority operated. Workers with considerable service moving to the 'new' container areas from declining upstream piers thus found themselves outranked for preferred work by new entrants on the latter's 'home' piers. Workers claimed income guarantees rather than move, and the result was escalating wage bills combined with labour shortage.

In the period before the 1971 strike, therefore, East and West Coast longshoring experienced different trajectories of change, the former associated with gradualism and conflict, the latter with a 'buy-out' of work rules to pave the way for new equipment. The basic difference between the two locations has been summarized by Goldberg, referring to the West Coast:

> The joint control by labor and management of the size of the registered labor force and of guarantees of employment have been of basic importance in promoting the acceptance of change. (1973, 267.)

Substantial differences in employment and earnings security do emerge. Whereas the registered work-force on the West Coast fell by less that 3.5 per cent between 1953 and 1966, the numerically much larger New York work-force declined by over 52 per cent (Levinson *et al.* 1971, 354, 525). Moreover, employment stability differed. Goldberg demonstrates that in 1965 only 7 per cent of hours on the West Coast were worked by casual labour, whereas 80 per cent of hours were contributed by permanent 'Class A' men. The Department of Labor study, however, found that, outside New York, 50 per cent of the East Coast work-force was casual (Goldberg 1973, 267; Levinson *et al.* 1971, 349). Whereas earnings equalization was pursued on the West Coast, substantial earnings differences remained in New York: in 1971, 6500 employees earned more than $10 000 p.a. but 3000 earned less than $3000 (Goldberg 1973, 270).

However, in one sense longshore work in the East retained a security which the West Coast did not have. The 1962 agreement on the East Coast established a 'container corridor' between Maine and Virginia extending up to 50 miles inland and within which all container packing was to be done by ILA labour. As Wilson notes, 'The rigid definition was drawn up before the main burst of container investment and thus the battle for job rights at inland depots was largely pre-empted' (1972, 144). The East Coast thus avoided the battles typical of the UK industry in 1972. It also managed to avoid the problems faced by the Devlin Report and on the West Coast. The M&M agreement, too, had ignored the exact definition of dockwork in giving *carte blanche* for new investment in exchange for job security.

These factors strongly influenced the approach to events in 1970 and 1971. On both coasts productivity had increased throughout the 1960s: between 1965 and 1970 Goldberg calculates an improvement of 55 per cent in the East and remarkably, since the base level was higher, by 68 per cent in the West.[27] However, the number of hours worked fell as the technological changes were implemented: in New York, hours worked fell from 42.1 million in 1964 to 32.8 million in 1970, while on the West Coast the fall was shorter and sharper from 24.3 million in 1969 to 19.7 million in 1970 (Goldberg 1973, 267–8). Earnings and job security thus came under threat, but different histories caused conflict over different issues.

The 1971 dispute began in the West on contract expiration on 1 July:[28] Wilson reports declining support for the M&M agreements from 1966 onwards, and the problem of declining job rights as container work, over which the M&M agreements gave no control, moved inland (1972, 146). The East Coast strike concerned guaranteed earnings: specifically, the refusal of employers to meet the rapidly escalating costs of guaranteeing the earnings of an immobile work-force while suffering labour shortage in container ports. Both disputes incurred Taft–Hartley injunctions: the West Coast dispute resumed after the injunction expired. Both settlements involved relatively high pay rises which subsequently fell foul of the then current incomes policy, revealing once more the low priority of pure labour-cost considerations compared to continuity of working.[29]

The West Coast settlement reached under threat of specific legislation in February 1972 improved earnings and job security and imposed a tax of $1 per ton on all containers stuffed or emptied within 50 miles of a port. Since the issue concerned movement of work inland, it also required resolution of a major jurisdication dispute with the Teamster' Union. The East Coast settlement was in some ways more radical, involving improved permanency of employment, the reduction in categories of labour, a radical reform of the seniority system, and a system of 'container royalty' payments where work was not performed by ILA members. In New York, overall control over labour relations—including hiring and technological change—was vested in a joint NYSA–ILA Contract Board.

Subsequently, the container issue did not simply disappear. On the East Coast, the NLRB objected to the rules of control of container work established in the 1972 settlement. This provoked a strike in 1977, and the issue was only resolved in 1980 as part of the first master contract on the Atlantic and Gulf coasts. On the West Coast, the pre-1971 pattern of relatively peaceful relations was re-established in agreements of 1975 and 1978; however, the 1981 contract included provisions—in the form of investment guarantees by the PMA—aimed at reversing the flow of container work away from the docks.[30]

The parallels between UK and US dockwork emerge clearly from the above discussion. Casualism, restrictive practices, conflict over the negotiation of change involving the breakdown of collective bargaining, and the involvement of government were all features of the East Coast scene. Experience on the West Coast, though different, reinforces the agreement that these connections are likely: the early decasualization and the pre-emptive offer of job security by employers before containerization secured a period of comparative peace but, as with the Devlin recommendations in the UK, the failure to foresee the movement of dockwork inland proved a stumbling-block.

Newspapers

Although differences in labour-market do exist between the UK and US newspaper industries, a number of basic features are shared. The UK industry remains more highly unionized, but a similar pattern of union fragmentation exists, and the NGA has a roughly similar craft-union counterpart in the International Typographical Union which historically has retained jurisdiction over composing-room functions though not, as the NGA does in the UK, over the machine room. Similarly, the ITU has controlled the supply of labour to newspapers, on the basis of seniority as in the UK:

> The union, not the management, decided whether piecework or time payment was to be offered, how the work was to be handed out during the day for the various groups of workers, how many trainees were to be admitted to each chapel. (Smith 1980, 209.)

In the USA, this system has also been associated with the development of practices to increase earnings and employment, such as machine-room sabotage, resetting of incoming set material, and hiring of extra workers irrespective of work-load (Zimbalist 1979, 118; Winsbury 1975, 20, 25; Smith 1980, 213). Against the background of an essentially similar product-market, casual attitudes have developed. However, although there is essentially a subcontractual system, the status of collective agreements in the USA concentrates bargaining at the end of a contract: the contractual form is thus closer to a contingent claims that a sequential spot contract.

The spread of photocomposing techniques in the USA has been far more rapid: in 1979, 84 per cent of all daily newspapers in the USA were photoset, although this did not necessarily imply a loss of jurisdiction by the ITU (Winsbury 1975, 52–7). By the end of the 1970s, some newspapers were considering 'second generation' systems (Smith 1980, 73–135). Whether progress in the UK provincial or national press is used as a yardstick—the distinction is inappropriate in the USA where the industry is predominantly localized—the USA had a technological 'lead' of several years.

The history of the ITU's attempts to control the advance of technology is by now well documented (Kelber and Schlesinger 1967; Zimbalist 1979; Smith 1980). A number of key differences from the UK emerge. The first is, of course, timing: the first strikes against the introduction of phototypesetting equipment took place in Chicago and New York as early as 1947 (Kelber and Schlesinger 1967, 30–40). The second is that the heavily unionized areas such as New York were unable fully to insulate themselves from the success achieved by employers in introducing new technology in poorly unionized regions such as Florida and Virginia (Winsbury 1975, 23–6; Martin 1981, 346): whereas Fleet Street has not noticeably been pressured into change by developments in the provincial press, union defeats elsewhere or innovation in the absence of unionization appear to have influenced developments in the crucial area of New York.[31]

Moreover, events in New York in the early 1970s occurred against a background very different from that in Fleet Street five years later. As Martin notes, UK moves towards new technology tended to blow hot and cold with the financial fortunes of the industry: when these picked up after 1975, the pressures for change became less intense. However, when the important negotiations between the ITU and the *New York Times* began in 1973, the industry had full knowledge of the consequences of failure to agree: two major disputes, in 1963 and 1966, had each been followed by the closure of several titles and resultant job loss. Although the ITU in New York managed to secure a veto over all new printing techniques in 1965, this was by then both anomalous in a national context and fragile given the financial state of the newspapers. Unlike in the UK, New York publishers managed to combine for subsequent negotiations as an employers' association.[32]

The agreement eventually reached in 1974 removed the veto on new technology at the three remaining New York papers—the *Post, Times,* and *Daily News*. Intended to last eleven years, it provided for lifetime employment for over 1600 ITU members, and for the retention of jurisdiction over certain new equipment for a five-year period. In return, the publishers secured unlimited rights to introduce not only automated typesetting but 'any technological device which [increased] the efficiency of their operation', together with a guarantee preventing the introduction of new make-work rules:[33] in fact, direct inputting was achieved in the *Post* and *Times* by 1978 (Rogers and Friedman 1980, 119).

The New York deal broke an 'impasse' in the introduction of new technology: soon afterwards two Washington papers, the *Star–News* and the *Post*, concluded a six-year contract granting lifetime job security in return for the unlimited right to introduce new photocomposing equipment; and in 1976 the ITU signed a ten-year contract with

New York commercial printing firms on similar terms.[34] The essentially cost-cutting nature of these changes, and a remarkable parallel to the Fleet Street experience, can be seen from the subsequent pattern of disputes. Just as cost-cutting exercises in the machine room occurred either with or instead of use of new technology at *The Times*, the *Observer*, and the *Financial Times*, so lengthy disputes occurred in the *Washington Post* in 1970 and in three New York papers in 1978 over crew sizes in the press-room.[35] Although compromise was achieved in New York following a break in the solidarity of employers, the pressmen's union at the *Washington Post* suffered decertification.[36]

Martin (1981) suggests four main reasons for the differences in outcomes of attempts at innovation between the UK and the USA. Firstly, the industry in the USA has been more dispersed, and competition between titles less intense. Secondly, trade union organization and solidarity were weaker in the USA; Griffin echoes this argument, noting the importance of NGA control over the machine room in the UK (1984, 41–6). Thirdly, the depth of computing experience and the reliability of systems were considerably greater in the USA. Fourthly, the employers combined effectively to negotiate change with unions on a regional basis: following the failure of the 'Programme for Action' they did not do so on Fleet Street.

Two further factors need to be added to this list. The first is simply that the UK unions were aware of previous experience in the USA, and the journalists, for example, resolved not to follow the policies of the Newspapers Guild in receiving disputed work (NUJ 1978), while the NGA could see in the falling membership of the ITU the costs of failure. The second is that market constraints and the status of collective contracts in the USA appear to have prevented the full development of a spot-bargaining system insulated from economic constraints on the Fleet Street model: as a consequence, though divisions emerged between, for example, the International and Local union over the 1974 negotiations in New York, collective bargaining, as a vehicle for change, did not suffer a breakdown similar to that in the UK in 1977.

CONCLUSION

In terms of the discussions of Chapter 4, we are dealing here with two different types of technical advance: on the one hand, the application of mechanization to dockwork involving the use of mass-production techniques and, on the other, the application of computerization to production processes. However, both changes were essentially process innovations of the cost-minimizing variety which impinged upon labour-markets operating sequential spot-contractual modes of almost Byzantine complexity: in both industries and both countries such forms of contracting could not survive the change. In both industries, product-market unpredictability provided labour with considerable uncertainty, but the requirement of rapid turn-around in dockwork and the perishability of the product in newspapers have enabled appropriately organized employees to exert substantial power in recurrent negotiations over wages, output, and employment.

The parallels between the two countries are perhaps more pronounced in dockwork than in newspapers. In both countries, dockwork has been characterized by substantial conflict over change, by the failure of established collective bargaining mechanisms to accommodate radical change, by the involvement of government agencies in independent inquiries and by the establishment of long-term high-cost guarantees of job security or lifetime earnings. Comparison of the newspaper industry in the two countries is less conclusive, in part simply because the outcome of change on Fleet Street is as yet unknown; however, craft resistance, high severance payments, and long-term employment guarantees have featured in both countries.

The broad comparison nevertheless conforms to the predictions of Chapter 5: spot contracting is difficult to eradicate but incompatible with high process efficiency. The problem of change in industries which experience the combination of cost-minimizing technological change and spot contracting is thus distinctive: the parallels may be extended both to Australia and to New Zealand (Griffin 1984). Other industries display markedly different patterns: one might, given space, effect similar comparisons between the UK and USA for the 'internal labour-market' industries such as railways and steel.

However, in both of the countries considered so far there was little variance between employers in labour relations strategy in that the same devices were adopted to cope with product-market volatility. Relative organizational efficiency was not tested by intra-industry competition. The case of motor vehicles, considered in the next chapter, thus differs substantially.

NOTES

1. For example, as late as 1965 there were 114 employers in Liverpool, 90 in Hull, and 77 in Bristol (Wilson 1972, 179).
2. In practice, a proportion of dockers were permanent. This percentage stood at 12.6 per cent in 1947 and 23.2 per cent on the eve of decasualization in 1967 (Wilson 1972, 308).
3. The continuity rule began during the war with the employer's insistence that the movement of cargo on and off ship should be completed by the men who started the job: it was intended to prevent employees shifting to better-paid assignments. It translated into a bar on mobility (Wilson 1972, 216–7; Jackson 1973, 82, 123).
4. Jackson expresses numbers 'proving attendance' (i.e. unable to find work but claiming maintenance) as a percentage of the register. Durcan *et al*. secure a higher percentage in general for this decade by expressing these numbers as a percentage of 'those available'.
5. Firms such as Olsen's in London used pallets in the 1960s with considerable efficiency improvements (see Wilson 1972, 241–3, 259–60); Mellish's 'firm B' is almost certainly Olsen's (See Mellish 1973, 103–23).
6. This was not new, but in the past the contingencies or 'state of the world conditions' as Williamson would have it, were negotiable.
7. These changes were not confined to London. Fadem (1976) notes them also in Southampton—decreased earnings variance and greater surveillance by foreman (138–9, 150)—and Grangemouth—improved mobility and timekeeping (182, 191)—but not in Glasgow (227 f.).
8. For accounts of these disputes, see: Wilson (1972, 146–54); Weekes *et al*. (1975, 278–94); and BJIR *Chronicle*, July 1975.
9. The BJIR *Chronicle* (1984) reports a further acceleration of job loss down to 14 000 in 1983.
10. These have been reduced somewhat by amalgamation. At the time of writing they are (in London) the NGA, AUEW, EETPU, and SOGAT.
11. In fact, there are different types of casual workers: see Sisson (1975, 72–6) and ACAS (1976, 32–4) for a discussion of stratification within the casual labour area.
12. ACAS presents essentially the same argument (1976, 27).
13. See, for example, the discussion of ownership of *The Times* in Martin (1981, 256–66). Martin argues that the relative unimportance financially of newspaper interests in the Thompson empire was significant, as was the fact that 'identification with British national interests was thought to be especially important for a foreign based company' (264).
14. Some newspapers, such as the *Mirror*, were at the time owned by groups with newsprint interests: see Martin (1981, 304–10).
15. For example, Winsbury (1975) the NGA newspaper *Print* (1971, various issues), and Royal Commission on the Press (1976, 46 f.).

16. See Martin (1981, 296–7) for estimates of projected productivity improvements at Times newspapers. For the view that there was a substantial productivity/quality trade-off, see NUJ (1978).
17. Quoted in Martin (1981, 97).
18. The following sections of Martin's work are relevant here: 198–200 Chs. 6–8. Other sources are quoted where relevant.
19. *Sic.* London Agreement between Mirror Group Newspapers Ltd. and the NGA, July 1977.
20. *Industrial Relations Review and Report*, No. 302, Aug. 1983; and No. 308, Nov. 1983.
21. The *Observer* experienced a dispute over such changes in the summer of 1980.
22. See *Industrial Relations Review and Report*, No. 271, May 1982; No. 294, Apr. 1983; and No. 318, Apr. 1984.
23. US longshore negotiations are characterized by a more direct involvement of shipowner interests, particularly in the West Coast coast wide bargaining arrangements. This is perhaps one reason for the earlier awareness of containerization.
24. The major rules deleted concerned sling loads, first place of rest, multiple handling, gang size, and manning scales (Levison *et al.* 1971, 323).
25. The following paragraphs rely on Levinson *et al*. (1971, 34–372), Goldberg (1973); and Jensen (1974) (except where specified).
26. The best description is that of Jensen (1974, 342–85).
27. In 1965, New York labour moved on average 0.434 tons per man-hour; the West Coast figure was 0.887 tons per man-hour (Goldberg 1973, 266).
28. The following account relies mainly on issues of the *Monthly Labor Review*, Sept. 1971 onwards.
29. See Goldberg (1973, 285): 'The ratio of (dock) labour costs to fixed costs, amounting to about 75 per cent under conventional systems, is only 25–50 per cent in the case of capital intensive container ship operations.' Both settlements were subsequently restricted by the Pay Board (see *Monthly Labor Review*, Vol. 95, Nos. 6 and 7, 1972).
30. See *Monthly Labor Review*, 97 (9), 62; 98(2), 86; 98(10), 62; 100(7), 52; 101(10), 56; 103(7), 570 104(9), 50.
31. In the 1960s the area was the 'printing capital of America' with one-fifth of all printing personnel employed (Smith 1980, 227).
32. This also occurred in Chicago (Martin 1981, 348).
33. *Monthly Labor Review*, Vol. 97, No. 8, 1974, 88–9.
34. *Monthly Labor Review*, Vol. 97, No. 11, 1974, 67–8.
35. See 'Why the Presses have stopped at the *Financial Times*', produced by journalists at the *Financial Times* (mimeo, n.d.). For the USA, see *Monthly Labor Review*, Vol. 99, No. 5, 1976, 52–3; Smith (1980, 231).

There was also a dispute over manpower reductions in several areas of the *Star* in 1978.

36. The agreements at New York were apparently facilitated by involvement of the Federal Mediation and Conciliation Service in an innovative way. A device known as 'medarts' (i.e. mediation/arbitration) was used which involved dealing with non-economic items separately from and prior to economic ones: all the former items were mediated to completion with each union separately, and the penalty for non-agreement was exclusion from discussion on economic issues. Crucial items might be independently arbitrated. (Interview, Director FMCS, 23 Apr. 1981.)

7

Fordism and Efficiency: the Car Industry

INTRODUCTION

THE car industry is dealt with separately because the pattern of strike activity over technological change has differed from that in dockwork and newspapers. Whereas conflict over change in those industries represented peaks in their strike records, such strikes in the car industry are a recurrent rather than episodic feature. For example, in the car industry, the average loss from strikes attributable to this cause in the 1970s was 44.7 thousand working days lost per annum, while the *lowest* working days lost total was over 3.2 million per annum (in 1976): technological strikes never accounted for more than 1.5 per cent of working days lost in a given year.[1]

The car industry thus features here primarily because of its size rather than because technical change has proved to be the major source of conflict. Nevertheless, the pattern of strikes over change does provide further support for the approach presented in Chapter 5, and some parallels with the two industries already discussed do emerge. In this chapter I shall try once more to show the link between spot contracting and conflict over change. The starting-point is the fact—discussed in more detail below—that strikes over change have been largely restricted to those parts of the industry which retained some form of spot contracting in the post-war period. Under such systems, the frequent product and process change in the industry generated recurrent conflict over payment for change. Elsewhere in the industry, different and more comprehensive forms of contracting produced very different patterns of conflict. The division is essentially between UK-owned and US-owned firms.

In fact, the argument extends beyond the experience of strike activity to a discussion of the efficacy of different contractual forms. The car industry is, *par excellence*, a mass-production industry to which, if the arguments of Chapters 4 and 5 are valid, comprehensive contracts most closely fit. I shall argue in this chapter that those firms which had the closest fit between process technology, product and labour contract were, on a number of indicators, most successful: they were also able to offer greater job security.

The structure of this chapter is as follows. Firstly, I shall describe the productive system, the pattern of unionization, and the idea of efficiency associated with car assembly. Secondly, I shall present evidence on the pattern of strike activity over change and relate it to different contractual forms. In the third section, I shall try to show that comprehensive contracting may be more efficient in this industry than spot contracting, and provide some comparative material on the USA in support of this contention.

PROCESS EFFICIENCY AND INNOVATION

Figure 7.1 outlines the basic sequence of operations in car manufacture. Beginning with the raw-material suppliers, two sets of operations proceed. The first has to do with the production of the power-train (essentially the engine, transmission, and axles),[2] the parts of which are first cast or forged, then subsequently machined and assembled. The second results in the production of a welded and painted car body: initially, sheet metal is stamped in a press shop to produce panels; these are welded to form subassemblies and then a rigid body in the body shop, and are finally painted. The final process, which marries the results of these two sets of operations, involves the installation of the thousands of components which constitute 'trim' and the fitting of the power-train to the trimmed body.

These operations need not all occur on the same site. In fact, car manufacturers differ in the extent to which they become involved in all of the stages of Figure 7.1: most produce their own engines, gearboxes, and panels, but collaborative ventures such as that between Austin Rover and both Volkswagen and Honda mean that basic components may be shared. Historically, the extent of vertical integration has varied both over time and between firms: Ford for example, have tended to integrate backwards into the production of glass and steel, but this has not resulted in a straightforward chronological acquisition of supplier companies, and different *stages* in the production process support different levels of backward integration (Abernathy 1978, 108, 139).

Finally, these operations differ in the extent of automation and in its timing. Press shops have been subjected to incremental improvement in the size and speed of operation, while, beginning in the

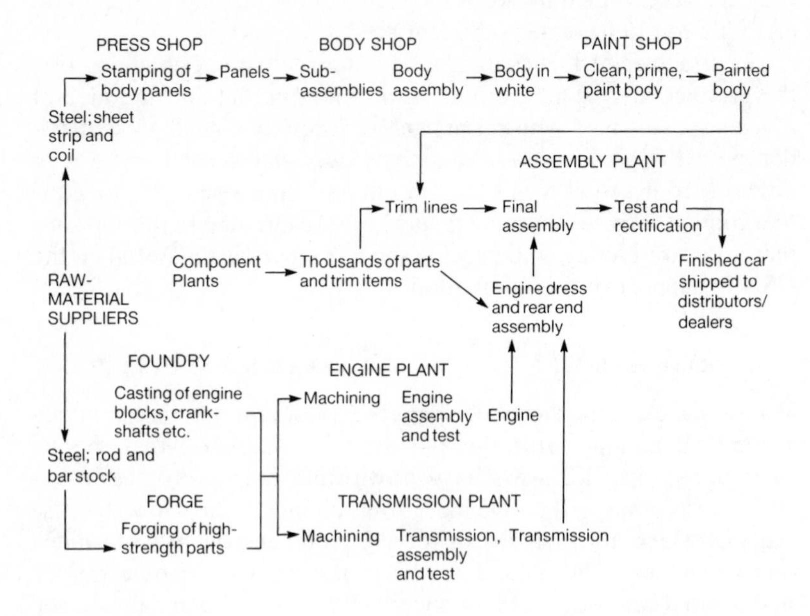

Figure 7.1. The Manufacturing Process in Car Production
Source: Hartley 1981

1950s, increasingly sophisticated automated transfer machinery has been introduced to operate between presses. Similarly, the late 1940s and 1950s saw the development of first-generation transfer lines in the engine-machining areas (Maxcy and Silberston 1959, 56–58; Ray 1969; Bloomfield 1978, 43). However, final assembly techniques have not changed a great deal since the introduction of moving assembly lines in the 1920s. Body assembly operations have been revolutionized in the 1980s with the advent in the UK of robotics and advanced automated techniques. I shall discuss this in more detail below. However, during the period currently under discussion, body and final assembly together accounted for roughly 50 per cent of labour costs (CPRS 1975, 14–16).

Different stages of the production process experience different economies of scale, although estimates of precise minimum efficient size also vary. One set of estimates is presented in Table 7.1. Overall, economies of scale have probably increased in the industry over time;

Table 7.1. Estimates of Minimum Efficient Scale in the Car Industry (000 units p.a.)

Source of estimates	Operations			
	Foundry and forge	Pressing	Engine and transmission	Final assembly
Pratten[a]	1000	500	250	300
Rhys[b]	200	2000	1000	200
White[c]	'small'	400	260	200
Ford UK[d]	2000			300
McGee[e]		2000		
Euro-Economics[f]	2000	2000	1000	250

[a] Pratten 1971.
[b] Rhys 1972.
[c] L. J. White, *The Automobile Industry Since 1945*, Cambridge, Mass., 1971.
[d] Evidence to the House of Commons Expenditure Committee Session 1974–75, Minutes of Evidence taken before the Trade and Industry Sub-Committee VoL I.
[e] J. S. McGee, 'Economies of Size in Auto Body Manufacture', *Journal of Law and Economics*, Vol. XVI (2), Oct. 1973.
[f] *Euro-Economics*, 'The European Car Industry—The Problem of Structure and Overcapacity', Mar. 1975.
Source: Owen 1983.

indeed, increasing the volume of output has always been an important way of reducing unit costs (Black 1980; Pratten 1971). This implies not only that output of a given model must be large, but also that car parts such as engines and gearboxes are used across a range of models.

It also places considerable emphasis on high capacity utilization to reduce unit costs, which in turn influences the preferred form of labour utilization. Some of the problems here can be considered with the help of Figure 7.2 which deals with the build-up of unit labour costs in car manufacture. Essentially, the figure consists of three elements: first, plant layout and automation, which determine capacity; secondly, labour-supply considerations, which influence the price of labour per hour; and thirdly, *lost* capacity, caused by the factors listed. As Rhys notes, companies of equal standard capacity will be differentiated on cost grounds by capacity utilization (1974, 15); and at higher levels of capital intensity, lost capacity increases in cost (Willman 1984, 5). On cost grounds, there are substantial pressures to reduce the differences between, on the one hand, capacity and output and, on the other, standard manning requirements and actual

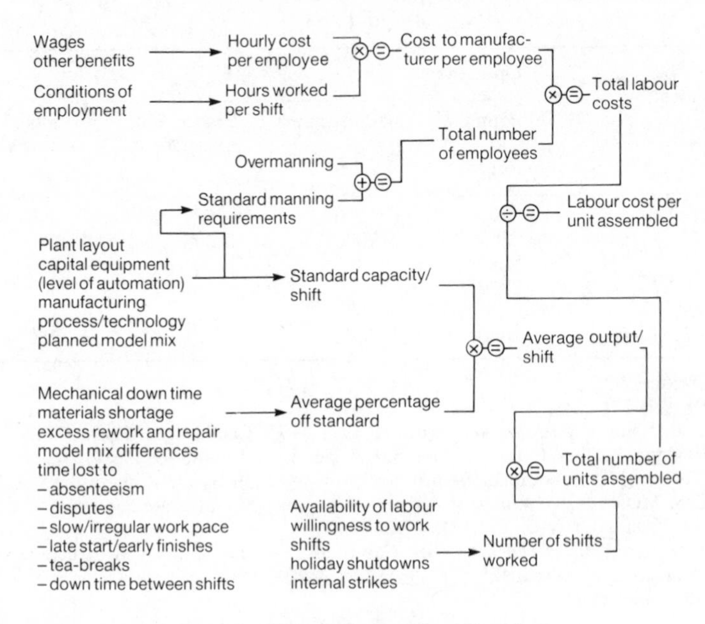

Figure 7.2. Unit Costs in Car Manufacture
Source: Hartley 1981

employment numbers. This focuses attention unequivocally on labour relations: the CPRS showed that disputes could account for up to 80 per cent of off-standard time in the UK in 1975; moreover, the greatest labour-cost disadvantages arose from low output and overmanning in assembly areas (CPRS 1975, 93, 97; see also Willman and Winch 1985, Ch. 5, 8).

The avoidance of inefficiencies of this sort implies a distinctive approach to labour relations. Effort bargains need to secure relatively high work intensity while the system of work organization which enhances process efficiency will economize on buffer stocks and wastage between work stations, as well as avoiding rectification work. To fit these demands, collective bargaining arrangements are under pressure not to introduce these kinds of inefficiencies through constant bargaining, nor to inflate labour costs (Williamson 1980, 1982).

However, the car industry is perhaps untypical in that economies of scale are not 'static'. Car manufacturers produce new models every four of five years, perhaps involving new technical advances in process and product. Since, historically, assembly lines were dedicated to one particular model, process and product change tended to go together. Competition, and the barriers to entry into the industry, have centred around the costs of model *change* (Abernathy 1978, 39–47; Black 1980, 256–60). The balance for car producers thus lies between efficiency in the exploitation of economies of scale on the one hand, and frequent product change across a range of models on the other. The history of the industry is dominated by the example of Ford, both in success and failure. The simple competitive strategy of low price, standardized design, and mass production of the Model T was extremely successful in the 1920s but ultimately gave way to General Motor's strategy of annual model change and a differentiated product. The basis of competition had shifted from cost to cost and performance. Ford's failure in product innovations caused permanent loss of US market leadership to GM (Abernathy and Wayne 1974).

This discussion of efficiency and innovativeness relates directly to effort-bargain issues. The strategy which developed the Model T also developed a distinctive labour strategy associated with moving assembly lines and mass production. 'Fordism', as it has been termed by several writers, was defined in terms of machine pacing of work, fragmentation of tasks, high levels of effort, and unusually high levels of reward (Sabel 1982; Lewchuk 1983). However, where products are differentiated and innovation is recurrent, pressures for some form of spot contracting at the point of change arise from employees. Employers may accede to this where product-market conditions are uncertain, even though process efficiency may be jeopardized. In short, in the post-war car industry the choice was between high-volume production and a comprehensive contract, or the transmission of product-market variations through to employees by spot contracts. A priori, one might expect the former to be less conflict-prone and more efficient than the latter.

TECHNICAL CHANGE AND STRIKES

This expectation is borne out by both the distribution and ostensible cause of stoppages over technical change in the UK industry between 1960 and 1980. Table 7.2 is compiled from the principal-stoppages list in the same way as before: the stoppages are detailed by cause, and by the company in which they occurred.[3] The main disputes are

Table 7.2. 'Technology Strikes' in the UK Car Industry, 1960–1980

Year	Company	Issue	Working days lost
1964	BL	Price for new car part	12 200
	BL	Price for new car part	6 900
1966	BL	Rate for new engine work	70 000
1967	BL	Rate for new body-fitting work	18 900
1969	BL	Demarcation over spot-welding equipment	18 700
1970	Vauxhall	Rate for operating new machinery	16 600
	BL	Rates for new model	111 700
1971	BL	Rates for new model parts	17 400
	BL	Relocation of equipment	12 000
	BL	Rates for new engine	52 300
1972	BL	Relocation of equipment	5 200
1973	Talbot	Demarcation	17 600
1974	Talbot	Demarcation	19 100
1975	Supplier	Pay for new equipment	13 100
1976	BL	Demarcation	28 900
1978	BL	Rates for automated equipment	139 400
1979	Supplier	Rates for new equipment	50 100

Source: DE *Gazette*.

over pay levels for new products until 1972, and thereafter demarcation disputes over the introduction of new equipment predominate. Only one dispute involved an American subsidiary: Ford are not involved at all.

These events are of interest, but the limitations of the data presented here must be stressed. These conflicts constituted only a small proportion of the working days lost in the industry, and companies such as Ford, which did not experience them, did not necessarily experience lower levels of conflict overall. Because many of the disputes were over piece-work prices, it seems reasonable to suggest that many similar but smaller conflicts occurred which are not accessible through the DE list. Moreover, disputes may have involved new technology but are not classified as having done so: this is the problem of attribution discussed in Chapter 3. Finally, strikes are only one form of resistance, and one could use the argument that plants which experienced strikes over change did not thereby suffer from restrictive practices.

Nevertheless, the distribution of such strikes towards those companies with a tradition of local spot bargaining is of substantial interest. Previous chapters have implied that such spot-bargaining systems may also support other forms of inefficiency and resistance to change.

It is thus worth looking more closely at the relationship between technical innovation and the effort bargain in these strike-prone companies.

INNOVATION AND PRODUCTIVITY

A brief summary of the history of technological change in the UK motor industry might simply say that, in general, process innovations diffused to the UK from the USA. This is perhaps best illustrated by example. Whereas diffusion of mass-production techniques in the USA occurred between 1905 and 1923, in the UK it occurred between 1925 and 1931. Moving assembly techniques were established by Morris in the mid-1920s (although Cowley was not in any sense an integrated system until 1934) and at Longbridge by Austin between 1922 and 1925 (Black 1980, 140, 219; Lewchuk 1983, 82–111). All-steel bodies, introduced in the USA by Dodge in 1923, were produced by Pressed Steel Fisher for Morris in 1927, but at a higher relative price than in the USA since Morris could not achieve economies of scale and produce them in-house (Abernathy *et al.* 1983, 161; Black 1980, 253). Transfer lines in engine machining were established in the early 1950s in the USA and between 1950 and 1960 in the UK (Turner *et al.* 1967, 78 f.; Ray 1969)[4], but the lead time on diffusion was later to expand again: highly automated body-framing lines were established by General Motors at Lordstown in 1970 and by British Leyland at Longbridge in 1980 (Willman and Winch 1985, 43–65).

The two cases are not, of course, wholly independent, primarily because both GM and Ford operated in and imported to the UK from an early stage. Standardization of parts appears to have been more difficult for the UK from the outset while the pattern of demand for cars—a small affluent market—coincided with high product differentiation and the retention of performance-maximizing competitive strategies amongst UK producers at a time when Ford in the USA was cutting both costs and prices (Black 1980, 165–72). In the 1920s, there were far more manufacturers in the UK than in the USA, with far lower average output; in addition, there were far more models—fifty-five in the UK in 1931 compared with thirty-two in the USA in 1930 (Black 1980, 252–61; Rhys 1972, 12–15).

The direct technology transfer which established Dagenham on the model of the River Rouge plant in 1932–3 enabled Ford to expand their market share in the 1930s on the basis of mass production, a small model range, and a high proportion of shared components; Vauxhall followed suit. More advanced production techniques caused severe problems for Austin and Morris, and forced the smaller British firms out of small-car production (Maxcy and Silberston 1959, 104–6). Some British firms retained a high number of models at low volumes: for example, Morris still produced ten models in 1946 compared with Austin's four and Ford's three (PEP 1950, 27, 130).

The history of the major indigenous UK car producers in the post-war period is one of repeated defensive mergers, forming BMC in 1952, and British Leyland Motor Corporation in 1968. However, such mergers were not always accompanied by the exploitation of potential economies of scale, particularly in engine and body manufacture, and criticisms of British Leyland in the 1970s still focused on poor product range, poor productivity, and a failure to invest in the new techniques. To what extent, then, can this innovation lag be attributed to labour relations problems? It has been established that strikes occurred in the 1960s and 1970s over innovation in the industry. Moreover, previous disputes have been documented at Triumph in 1956 concerning the introduction of transfer machining on engine blocks and at Cowley in 1934 as the production system was being integrated (Melman 1958; Edwards 1979). Several analysts have argued that problematic labour relations have prevented the achievement of economies of scale in the UK industry, and one might plausibly suggest that such a failure might frustrate innovation which required high output levels (Jones and Prais 1978; Prais 1981, 145–63).[5] Finally, the CPRS have suggested that output in the UK industry did not match that on similar equipment abroad in the early 1970s and that work-pace tended to be slower on similar equipment in the UK than on the Continent (CPRS 1975, 96–105). Several different types of labour resistance are implied.

While productivity differences between UK car operations and those of the USA (and indeed elsewhere) are relatively well documented, and the charge that the industry has been under-capitalized because of labour relations problems has been made from several quarters, there do seem to be certain attribution problems. For example, Figure 7.3 shows productivity per man in the UK and USA car industries since 1920. The figures are derived from Black's calculations from USA and UK census returns. The ratio of US/UK

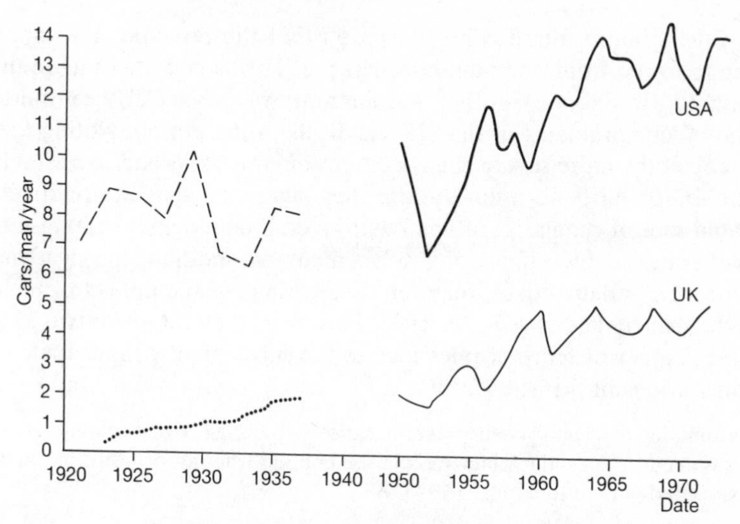

Figure 7.3. Productivity in US and UK Motor Vehicles, 1920–1970
Source: Black 1980

productivity per head has, in fact, declined over the period, although the absolute productivity differential has increased. However, the sequence strongly suggests that labour relations are not terribly important as an explanation of comparative productivity movements: the productivity ratio for the period 1923–7, when both industries were largely non-unionized, averages 7.24 : 1 whereas for the post-war unionized period it averages 3.14 : 1. Moreover, as I shall show below, the character of trade union organization in the UK changed markedly during the latter period. Black's calculations show that both sets of productivity figures are in fact extremely sensitive to variations in overall output, and the suggestion that economies of scale were more important in the explanation of productivity differences than labour relations receives support from his comparison of productivity at Ford's River Rouge and Dagenham plants. The latter in the 1930s was an identical but scaled-down version of the former; in 1934, both were non-unionized, but direct labour hours per car at River Rouge were only 57 per cent of those at Dagenham (Black 1980, 222). The comparisons of productivity differences made by the CPRS within Ford in 1975 thus have a very long history indeed, and it does seem that the 'productivity problem' in the UK car industry *pre-dates* the 'union problem'.

Furthermore, there is no indication that the post-war innovation lag followed from trade union resistance. Turner *et al.*, writing in the mid-1960s, did not feel that 'automation' was particularly central to the labour problems of the UK car firms in the period. Noting that many of the more severe changes to production skills had occurred in the 1920s, prior to unionization, they suggested that the relatively rapid rate of change, combined with overall employment expansion, had softened the effects of displacement; in addition, the post-war automation debate had, they felt, overestimated the impact of technological change (1967, 72–102). However, they do make two general points which are of relevance to the subsequently rapid spate of innovations in the 1980s. Firstly,

> automation has only been a direct cause of conflict where it has led to redundancy: and redundancy has caused conflict whether or not it has been associated with automation. (1967, 83.)

Secondly, a

> central problem appears to be that of finding a wage system appropriate to highly mechanised—but far from fully automated—technology. (1967, 97.)

Lack of innovation, a failure to achieve scale economies in model proliferation, and low productivity had thus been relatively enduring features of the UK car industry. Whereas, in the USA, companies quickly moved to oligopoly in which scale economies were extremely important, UK companies—the future components of BL—remained trapped in a small market characterized by high product differentiation. Concentration of production and rationalization of operations proceeded slowly. The relevance of this early history lies primarily in the fact that, faced with such a product-market, UK companies developed a relatively idiosyncratic set of institutions for labour management which were to become wholly inappropriate for high-volume production. In particular, these institutions acted as a brake on process efficiency and encouraged opposition to change.

The cyclical nature of the UK industry is illustrated in the trend of passenger-car output for the post-war period given in Figure 7.4. Passenger cars are the dominant rather than sole output measure for the UK vehicle industry, with commercial vehicles and components also to be taken into account. Nevertheless, the trend measure of output, which is itself a smoothed measure of quarterly or monthly variations and of changes at firm level, appears quite closely related

Figure 7.4. Production and Employment in UK Motor Vehicles, 1946–1982
Source: Production data, *Economic Trends*; employment data, DE *Gazette*

to employment trends up to 1973. The lack of employment security is illustrated by the fact that numbers employed dropped in thirteen of the thirty years under consideration.[6]

The sensitivity of employment trends to output variation actually signals the more pervasive influence of product-market developments on labour relations. The employment trend itself masks output-based variations in hours of work: this has been documented independently for the period 1946–62 by Turner *et al*. (1967, 161) and for 1951–73 by Durcan *et al*. (1983, 322). The latter authors also illustrate substantial regional shifts masked by the trend of Figure 7.4: in fact, 'the dispersion of employment worsened the irregularity of employment in those areas where production had previously been concentrated' (1983, 329). In addition, this variation caused substantial change in earnings, due to variations in overtime and piece-work earnings. Turner *et al*. show substantial weekly variation (1967, 163); Durcan

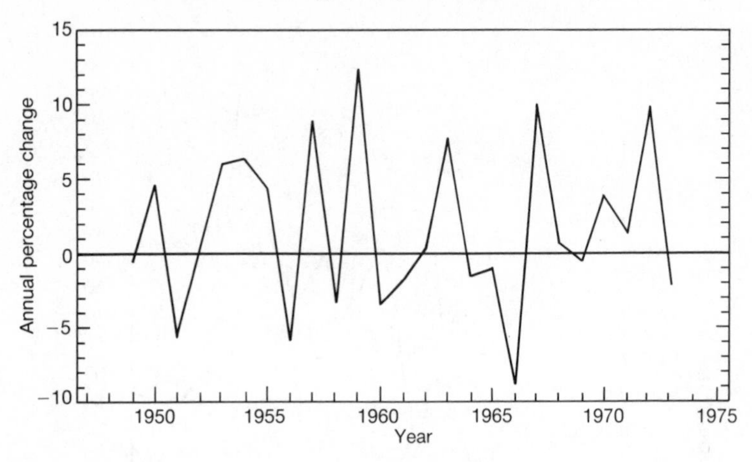

Figure 7.5. Changes in Net Disposable Earnings in UK Motor Vehicles, 1949–1973
Source: Durcan *et al.* 1983

et al.'s analysis, presented here as Figure 7.5, shows the longer-term fluctuations in net disposable earnings.

Irregularity of output is central to Turner *et al.*'s explanation of strike activity in the industry. Conflict was seen to arise from irregularities of earnings—particularly as there were marked anomalies between groups of similar workers in the industry—and of employment, but the tendency for working days lost to rise during recessions in output supported the broad suggestion that industrial disputes functioned as a means of dealing with short-term excesses in the supply of labour without recourse to redundancies or short-time working of the sort which had proved so conducive to conflict in the recession of 1956 (Turner *et al.* 1967, 104–28). However, Turner *et al.* also laid considerable emphasis on two sets of worker expectations and on the inability of institutions to deal with problems when they arose. Contrasting the pattern of inter-war and post-war labour disputes, they argued that workers developed over this period the 'fair wage' expectation—the idea that earnings from a particular job should be fair relative to other earnings—and the idea that performance in a job establishes job property rights. The *implications* of these notions are of some interest.

It is an obvious further implication of the 'fair wages' idea that earnings should be maintained by [sic] comparison with their recipients' past experience: and

> this has a particular reference to many piecework disputes in the industry. But the demand that work-loads, as well as wage rates, should be negotiable is also a logical extension of it, since it would be meaningless to insist that earnings should be 'fair' in comparative terms without requiring that efforts, too, be regulated by agreebly comparative standards . . .
>
> The concept of 'job property rights' . . . also includes the idea of rights to a particular job at a particular place, and may extend to the right to consultation in anything which may affect the future value of his (i.e. the manual worker's 'property'). (1967, 337.)

Employees therefore sought to bargain both about the level of input to the effort bargain and about its worth; they also sought some influence over its rate of change. These conceptions were developed in the face of a policy on the part of employers to accommodate product-market uncertainties by allowing variations in earnings and employment. At this point it is necessary to describe the development of the system—which in parts comes close to that described in the previous chapter in docks and print—and to contrast it with the preferred institutional solution of foreign-owned manufacturers, principally Ford.

Recent work by labour historians has established the pre-war origins of the system. Lewchuk, in particular, has attempted to show the differences between the labour relations strategies of UK employers and the high day-wage/high effort system established by Ford at the Trafford Park works in 1914. His explanation relies not on product-market developments, which he does not consider, but on the requirement to accommodate post-war labour antagonism through the use of laxer supervision and piece-work. At Morris Motors, there was in the 1920s some machine pacing, but also payment by results based on individuals or small groups; at Austin during the same period, piece-work existed without machine pacing (1983, 82–107). At Daimler, employees forced the establishment of a 'gang' system which retained elements of internal contracting: the employer was forced to accept a bonus-based system in which both wages and work allocation were the responsibility of the gang (Black 1980, 184). Similarly, sheet-metal workers throughout the industry operated systems of work pooling and pay equalization through shop stewards where possible (Tolliday 1986, 21). By contrast, piece-work was avoided in favour of machine-paced high day-rates with an 'efficiency' supplement at Vauxhall, and Ford simply operated the high day-rate strategy developed in the USA.

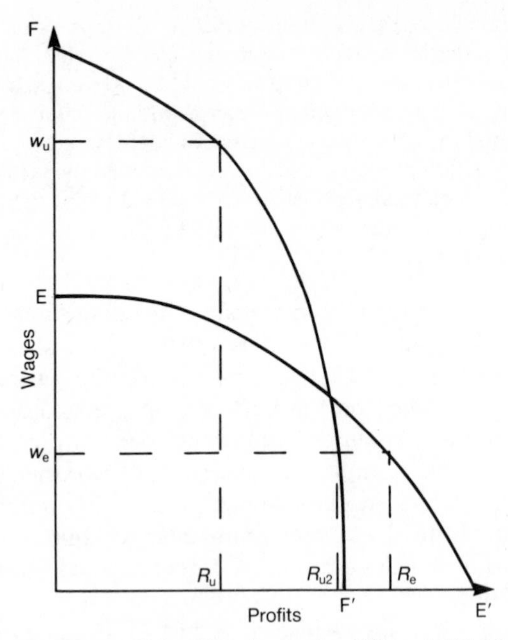

Figure 7.6. The Factor Price Frontier in the UK and US Motor Vehicle Industries
Source: Black 1980

Several authors have argued that this product and labour strategy made good sense in pre-war Britain. For example, Rhys argues that the higher capital/labour ratios used by US companies at that stage may have effectively raised the necessary break-even point, so that higher volumes were necessary to cover costs: in the UK, this meant very high levels of market penetration (Rhys 1972, 305). Black's argument is presented in Figure 7.6. He argues that an American technique FF′, applied under conditions of wages w_u and profits r_u may be less efficient where the wage/profit conditions are more appropriate to the UK technology EE′ (i.e. w_e, r_e apply). Under such conditions the English technique is superior to the American on account of the higher rate of profit ($r_e > r_{u2}$): Maxcy and Silberston are able to show that, before the war at least, profits at Ford and Vauxhall were inferior to those of Austin and Morris (1959, 233–37).[7]

However, product- and labour-market conditions changed substantially in the post-war period. Output of passenger cars increased threefold and hourly employment by 50 per cent between 1930 and 1950 (Black 1980, 157). Moreover, whereas before the war many car firms were non-union, after the war the situation had changed. While much of the car industry was unionized during the war, the system described by Turner *et al*., in which workplace organization and the power of the steward were well established, probably did not emerge until the late 1950s (Zeitlin 1980; Lyddon 1983). Longbridge and Cowley, for example, appear to have been weakly organized in 1956, though 100 per cent unionism existed in the car firms by the mid-1960s (Tolliday 1986; Turner *et al*. 1967).

In many companies the deterioration of piece-work schemes into work-force-controlled sequential spot contracts appears to have proceeded: the outcome was in some cases very similar to the sorts of contracting described in docks and newspapers. The, probably extreme, example of Standard Motors revealed union hiring, control of wages and work allocation, and mutality in the 1950s. At the same time, at Rootes, hiring was through the union office and stewards had apparently effective control of labour loading and the speed of the track (Tolliday 1986). Many of the component elements of the future BL and Chrysler thus displayed in varying degree the features of institutional obsolescence—the 'two systems' of industrial relations of the Donovan Report. In the terms employed here, they had opted for spot contracting.

A very different approach was adopted at Ford. Under duress, the company had recognized trade unions at the end of the Second World War (Beynon 1973, 43–7). From the outset, there was an insistence on centralized bargaining with national union officials, a refusal to accept the activities of shop stewards, and an insistence on management's right to manage on the shop-floor, including the exclusion of work-loads from negotiations. It is clear that negotiations did occur on the shop-floor at Ford, but equally clear that the degree of managerial control remained greater than that in other, federated, firms (Beynon 1973; Tolliday 1986). Ford never used piece-work and operated a company-wide grading system from the 1950s.

Although the company could not isolate itself from the rising expectations concerning wages and job rights which Turner *et al*. suggest influenced car-company employees in the post-war period, it experienced them rather differently. While operating in the UK market, Ford UK has increasingly been insulated from local fluctuations by the use of tied imports: its UK production capacity has for some

Table 7.3. Output and Employment in Ford UK and UK Motor Vehicles, 1949–1982

	Constant *a*	Production[b] *b*	R^2	Durbin Watson
Ford[a]	464.4	0.098	0.743	0.72
UK	184.0	0.143	0.861	0.95

Sources:
[a] Employment data from the DE *Gazette* and Ford Motor Company.
[b] Output data from the SMMT and Ford Motor Company.

time been lower than its market share; by 1982, Ford were importing almost 50 per cent of UK sales, a volume greater than their exports from the UK.[8] In addition, the company was the most profitable car operation in the UK in the post-war period (Maxcy and Silberston 1959).

Both of these factors enhanced Ford's ability to insulate employees from product-market fluctuations. The flat-rate basis of earnings at the company left overtime earnings as the principal variable component: as a result, earnings at the company have tended to vary less than those of the industry as a whole. In addition, there is some evidence to suggest that employment levels have varied rather less in the face of product-market change. Table 7.3 shows the relationship between output and employment for Ford UK and for the industry as a whole over the period 1949–82. The R^2 values indicate the extent to which changes in employment depend on output changes, and the slope coefficients (b) indicate the strength of the movement. Although this is a very crude indicator since it does not cater for model or price changes, it does imply a looser relationship between employment and output changes at Ford than elsewhere, and a smaller change in employment per unit of output.[9]

The institutional backdrop to this was a contingent claims contract. Ford favours company-wide comprehensive annually negotiated pay deals, with labour relations between bargaining dates being regulated by a 'Blue Book' of rules and procedures under a regime of managerial prerogative. Until 1980, a low-trust comprehensive contract had thus accompanied employment and earnings stability greater than that associated with UK competitors. Indeed, as Figure 7.7 indicates, throughout the difficult years of the 1970s, when employment fluctuations in the car industry were severe, Ford contracted employment only around the first oil crisis.

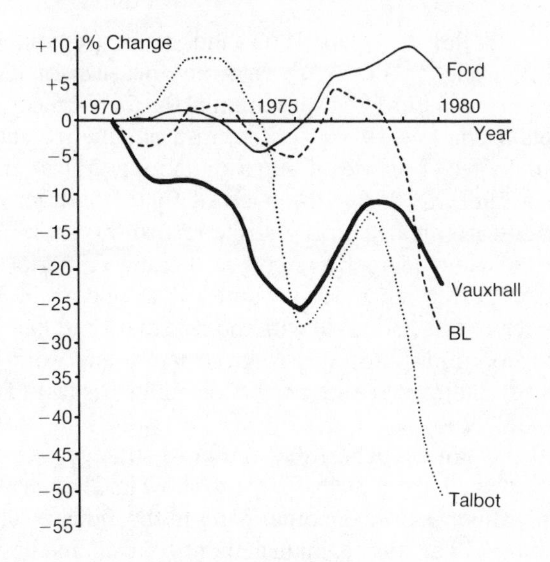

Figure 7.7. Percentage Change in Employment in UK Car Companies, 1971–1980
Sources: Company data from Annual Reports

Given these contractual differences, it is of some interest to note that the system used by Ford in the UK appears to have had superior efficiency qualities to that operated by home-owned UK firms. The evidence for this is circumstantial, but extends over a long period.

In the post-war period, Ford profits have consistently been the best of the UK producers (Maxcy and Silberston 1959, Appendix D; Young and Hood 1977, 169). In the 1970s, in particular, Ford avoided the substantial losses experienced by the other UK producers, remaining consistently in profit (Marsden *et al*. 1985): throughout the period, the company's market share expanded.[10] One could not attribute this mainly to labour relations issues, but a number of more direct measures seem to imply that, whatever the cause of Ford's relative success, the labour relations system was in rather better shape than those of the UK firms for much of the industry's history. Most measures of labour productivity show Ford UK in a favourable light in comparison with UK competitors, if not with overseas. For example, Black shows that direct labour productivity in Ford UK was much better than that in the UK firms at the same time as the unfavourable comparison with River Rouge (1980, 224). On a rough cars-per-head basis, Ford UK productivity was again the best of a bad UK lot between 1965 and 1974 (Young and Hood 1977, 157), while on the more rigorous measure of direct-labour man-hours per car, Ford UK substantially out-performed comparable BL models in the late 1970s, not because of superior equipment or design but largely because of superior labour utilization (Willman and Winch 1985, 150). On the basis of figures extracted from the company's own estimates, Ford's strike record was rather better than those of its main competitors. Over the entire period 1946–64, Ford's strike performance was far better than that of the BMC companies: much of this had to do with the fact that Ford had a relatively low proportion of disputes over wage structure and work-loads ('fair wage' issues) and over redundancy ('job rights' issues) (Turner *et al.* 1967, 73, 235, 260).

All of this is not to suggest that the Ford strategy was infallible or even the best available; but, as I hope to show in Chapter 9, it exerted a powerful influence over the other firms in the industry which sought centralization of bargaining, management control, and high day-rates in the 1970s.[11] The Ford labour relations strategy diffused across the UK industry, as had moving assembly techniques a generation earlier.

During the period 1955–75, a number of cost-cutting process innovations were adopted in the UK, but the volatility of the product-market was not conducive to massive capital investment. Hence 'automation' or the state of technology plays little part in Turner *et al.*'s discussion of the industry, and the emphasis is thus rather on the institutional blockages to 'the embodiment of new labour expectations into negotiated employment contracts', (1967, 343). Nevertheless, over the period, Ford remained a more capital-intensive operation than its UK competitors.

I have shown above that Ford preferred a more capital-intensive strategy before World War II: precise figures of fixed capital per worker are supplied by Lewcuk (1984, 24). This difference persisted in the post-war period. Maxcy and Silberston show Ford's net assets to be greater than those of BMC in 1956 despite the latter's 12 per cent market-share advantage (1959, 117, 178).[12] Silberston later showed that, whereas Ford's net assets grew by 73.6 per cent between 1954 and 1963, those of BMC grew by only 47.5 per cent (1965, 276). Finally, the House of Commons Expenditure Committee Report showed fixed assets per man at BMC in 1974 to be the lowest of the fifteen major European operations surveyed, including all three of the UK-based American operations (House of Commons 1975).

Without offering implausible single-factor explanations for investment patterns, one can suggest that Ford appeared to have a comparative innovative advantage within its labour relations system or, at least, that it did not suffer the disadvantages of the UK producers. Some of the problems involved here were summarized by a BL Longbridge shop steward in evidence to the Expenditure Committee: referring to the period before the introduction of measured daywork in the early 1970s, he argued that

> in piecework, because of the antiquated machinery, if anything happened to that machine that affected their livelihood, then the people who were brought up on piecework would put it right if they thought that no-one was looking at them—the skilled men or people like that—and they would keep the machine going with a piece of string. (Expenditure Committee, *Minutes of Evidence*, Vol. 1, 1975, 358.)

The contractual system thus harnessed employees' capacity in such a way as to provide local investment disincentives.

In summary, then, it is likely that Ford suffered fewer instances of the four types of obstruction outlined in Chapter 1: fewer strikes, no spot bargaining over change, fewer restrictive practices, and no evident cost increases. Overall I hope I have shown that the comprehensive contracting approach favoured by Ford was in fact more efficient than spot contracting where the pace of innovation was slow. The latter system, retained by the UK companies, was associated with low volumes, low productivity, and a range of labour relations problems.

THE USA

Despite the marked differences in legislative, institutional, and economic context, there is some support for the contention that Ford strategies in the UK were similar to those current in the USA car industry. For example, the system of collective bargaining operative in the USA, at least until 1979, avoided local pay bargaining or piece-work and relied on company-level formula-like wage-setting mechanisms which provided welcome continuity in labour costs in the light of the long lead times on new investment planning; codified job classifications, seniority arrangements, and management-rights clauses at local level; and provided 'connective' relations between

Table 7.4. Output and Employment in the US Car Industry, 1949–1973

	Constant *a*	Production *b*	R^2	Durbin Watson
USA	419.0	0.023	0.26	0.68

Source: Data from Black (1980).

plant and company level premised on contractual resolution of disagreements (Macdonald 1963; Katz 1985).

Moreover, while the US product-market was not immune from cyclical fluctuations, there is evidence to support the contention that this did not have the same effect as in the UK on employment and earnings. Table 7.4 repeats the calculations of Table 7.3, this time for the US industry. The loose relationship between employment and output, and the relatively small employment change per unit of output, are surprising given that Black's data refer only to production operatives: one might have expected higher variance within this category than in the work-force as a whole, and these figures might thus understate the degree of employment stability in comparison to the UK. These figures, of course, go only to 1973, but a similar calculations by Streeck and Hoff (1983) for the period 1970–80 confirm that employment change per unit of output was greater in the UK than in the USA: in fact, employment variation in the UK was the highest of seven countries surveyed.[13]

To some extent, these differences are a function of productivity differences, since employment variations are likely to be greatest where output per man is lowest. They may, nevertheless, indicate real differences between the UK and the USA in the uncertainty faced by car workers in the post-war period—which may have been enhanced by differences in contractual arrangements. In the USA, detailed contractual arrangements had developed to guide adjustments in employment and working hours: the core elements included seniority-based lay-off systems, inter-plant transfer rights, use of temporary part-time workers, and, since 1955, supplementary unemployment benefit (Katz and Karl 1983). In fact, writers comparing the car industry in the two countries since the war have argued that this contingent claims contract amounted to a more substantial and secure basis of local union control than that achieved in the UK through shop-steward control of spot-contracting arrangements (Tolliday and Zeitlin 1986).

Wage-fixing mechanisms provided earnings stability in the USA from 1948 onwards. Neither wages nor negotiated fringe benefits tended to vary with output in the short term. In fact, over a long period of time there were guarantees of improvements in real earnings through the operation of what Katz terms 'wage rules', i.e. formula-like mechanisms providing for standardized annual cost-of-living adjustments (COLA) and an annual improvement factor (AIF). Katz is able to demonstrate that these rules caused car workers' earnings to move with inflation *and* productivity improvements combined over the period 1948–81. They also evened out wages across companies and encouraged the eradication of piece-work (1985, 26–41).

The AIF is of particular interest since its explicit intent was to give employees a share in the benefit of technical progress. Katz quotes the following from a 1982 GM agreement:

> The improvement factor provided herein recognises that a continuing improvement in the standard of living of employees depends upon technological progress, better tools, methods, processes and equipment, and a cooperative attitude of all parties in such progress. (1985, 64.)

In the terms discussed above, the wage rules are thus a contingent claims contract of considerable complexity. Management reserves the right to act and the union to protest within detailed legalistic procedures. It was essentially a low-trust arrangement, but unions were bought off the use of techniques which might impede efficiency, and the companies preserved the freedom to make production decisions. As Katz notes, it was premised on a set of product- and labour-market conditions—and upon particular process technologies—which were broadly stable and have subsequently disappeared. Nevertheless, the industry avoided the conflicts over change experienced in the UK.

CONCLUSION

In this chapter and the previous one, I have looked at three UK industries in which there appeared to be resistance to technological change in the last twenty years or so. Strikingly, in all, forms of labour contracting involved recurrent bargaining which was incompatible with changes designed to improve the efficiency of mass-production techniques. In dockwork, the change was in effect the first advent of such techniques. In newspapers, the pressure arose out of a concern with labour and materials costs in the face of competition. In cars, recurrent innovation along with the retention of recurrent contracting yielded a pattern of conflict sustained over a long period of time.

In all the industries, resistance to process changes had its origins in the employer's prior decision to transmit product-market fluctuations through to employees. At the risk of overstatement, this decision actually *creates* a form of trade-unionism in which the responsibility for security of employment and earnings is devolved to shop stewards who cannot take efficiency and innovativeness into account. Similarly, contractual change in effect destroys this form of unionism.

The docks and newspapers examples reveal that there is nothing specifically British about that. In the car industry, where contractual differences exist within the UK and between the UK and USA, differences in efficiency and in employment and earnings security are in the directions which the approach of Chapter 5 would predict. Overall, the links between product and process change, on the one hand, and labour relations, on the other, are demonstrated. Spot-contracting forms are inefficient in mass-production industries, and are vulnerable to cost-minimizing pressures. As product-markets evolve, processes change and contractual forms must adapt.

However, such an approach implies that, once established, efficient processes and comprehensive contracts remain undisturbed. This can only be sustained by looking backwards rather than forwards, since certain types of change which affect a range of industries may disturb such relationships substantially. One such major change is the advent of micro-electronics.

NOTES

1. Since the figures on technological disputes are calculated from the list of principal stoppages, whereas the figures for the industry as a whole cover stoppages of all sizes, this figure is likely to be an underestimate. However, even if it is assumed that the same proportion of stoppages omitted from the principal-stoppages list are attributable to technological change as those included, the figure never rises above 3 per cent.
2. In Abernathy's definition, the power-train 'includes the engine, transmission, clutch, drive-shaft differential and axle. The term refers to the mechanical components that generate power and transmit it to the driving wheels' (1978, 12).
3. Simplification is involved here. The company names are those currently

in force, while component firms are aggregated under the heading 'Suppliers'.

4. An exception here was the very early establishment of partial transfer lines at Morris's Coventry Engines Plant in 1923 (Woollard 1954, 26). Turner *et al*. suggest that they were discarded as unreliable (1967, 78).
5. Moreover, the failure to achieve adequate size apparently occurred in 1960–75 when other sources saw labour relations in the industry as a problem (Prais 1981, 151–2; Turner *et al*. 1967).
6. Discontinuities in the employment series indicate changes in the SIC. As Durcan *et al*. note, the net effect of such changes, particularly that of 1958, is to increase the employment base. Hence this sequence cannot be used for the calculation of productivity trends as in Figure 7.3 (1983, 325).
7. Average net profits (before tax, plus other income, less depreciation, directors' salaries, interest, and before deduction of preference and ordinary dividends) were as follows for the period 1929–38:

Firm	£(000s)
Austin	708.3
Morris	1195.1
Ford	515.3
Vauxhall	512.7

Source: Maxcy and Silberston 1959.

8. SMMT figures. Ford Cortinas are manufactured at Ghent in Belgium, Escorts at Saarlouis in West Germany, and Fiestas at Valencia in Spain.
9. At Vauxhall, similar conditions applied, particularly after the renewal of bonus payments in 1957. For the period 1957–81, a calculation similar to that described in the text yields $R^2 = 0.46$, $b = 0.052$.
10. Ford's share increased from 26.5 per cent of new registrations in 1970 to 30.7 per cent in 1980 (SMMT).
11. Lewchuk (1984) argues that the Ford 'high day-rate' wage differential over the pay of UK manufacturers disappeared in the late 1930s and that this was linked to the rise of trade-unionism in the company. He also suggests that employees in the period after the First World War may have preferred high day-rates to piece-work, and that EEF policy was against it.
12. The same figures show fixed assets to be higher at Vauxhall than at Standard, on similar volumes.
13. Their figures for 1970–80 are: USA, $R^2 = 0.89$, $b = 0.41$; UK, $R^2 = 0.80$, $b = 0.89$ (Streeck and Hoff 1983).

8

Micro-electronics and Mature Industries

INTRODUCTION

THE examples of change discussed so far largely pre-date the impact of microprocessor technology in the UK, yet the threat posed by this technology triggered the second of the automation debates discussed in Chapter 2. It is thus appropriate to turn to a separate consideration of the impact of microprocessor technology and the response of trade unions in specific industries.

Micro-electronic technology is not simply an innovation; it is a major component in what Freeman *et al.* have termed a 'new technology system', the essential property of which is that it is associated with both the growth of new industries and with a widespread diffusion effect upon existing industries. It thus involves process and product innovation, but the mix between the two is likely to vary substantially. While new 'performance-maximizing' industries may arise, and microprocessors may be incorporated as improvements to existing products, some microprocessor applications may involve cost-minimizing process innovations in mature industries. Moreover, the application of such innovations extends outside manufacturing, particularly with the development of office automation.

The rate of diffusion, the balance of process and product innovation, and the economic circumstances of the receiving industries are all important for industrial relations. In particular, process innovations in mature industries are likely to be of considerable concern. In this chapter I shall briefly spell out the likely implications of microprocessor-based technology for industrial organizations, using the approach developed in Chapters 4 and 5.

THE DIFFUSION OF MICRO-ELECTRONICS

Freeman *et al.* suggest that microprocessor technology constitutes a 'new technology system'. Such systems involve related families of innovation occurring almost simultaneously. They tend to involve 'distinct new groupings of firms with their own "sub-culture" and distinct new technology and . . . new patterns of consumer behaviour' (1982, 64, 68). Such systems may be the basis of 'long waves' of economic development affecting old and generating new industries. A pattern of diffusion is thus implied.

Freeman's suggested diffusion path for micro-electronic technology is presented in Table 8.1: a comparison of this with Table 3.2 is of some interest. Although the fit is not perfect, it is clear that performance-maximizing activities cluster on the left of the table (i.e. are rapid adopters), while cost-minimizers cluster to the right. The table does more than merely emphasize the pervasiveness of micro-electronics, it also emphasizes the unevenness of diffusion: the flow has tended to be from the electronics industry itself, first to industries such as scientific instruments, where electronic subsystems represent a large proportion of total product cost, then to firms operating flow production systems for process control, and finally, in Freeman *et al.*'s view, to 'the older and slower growing (or declining) sectors of the economy with a low endowment of qualified engineers' (1982, 123–4).

The implications for different sectors are thus to some extent to reinforce existing patterns. Performance-maximizers will expand in number through growth of new industries and incorporate micro-electronics into products where possible. Sales-maximizers will engage in product development but will also use micro-electronics for process innovations in product design and production. Later stages of diffusion 'focus technical effort increasingly on cost reducing process innovations rather than new products'. (Freeman *et al.* 1982, 79).

This pattern is largely borne out by recent survey research on microprocessor applications in manufacturing industry. Northcott and Rogers' (1984) basic findings on the industrial distribution of product and process innovations reveal a concentration of product innovation in performance-maximizing and the biggest process users in sales maximizing or cost-minimizing industries (see Figure 8.1). Moreover, whereas product users saw advantages in better product performance (70 per cent), flexibility in product development (66 per cent), and consistent product quality (54 per cent), process users saw advantages in better process control (75 per cent), consistency of product (74 per cent), and more efficient use of labour (66 per cent). (Northcote *et al.* 1982.)

In terms of the use of different types of micro-electronics-based equipment in manufacturing production processes, it appears from

Table 8.1. Diffusion of Micro-electronic Applications

Rate of diffusion[a]	Rapid (from 1960)	Medium (from 1965)		Slow (from 1965)		
(Depth of impact)[b]	High	High	Medium	High	Medium	Low
Design and redesign of products to use micro-electronic technology	Electronic capital goods; Military and space equipment; some electronic consumer goods	Machine tools; vehicles; electronic consumer goods; instruments; some toys	Other consumer durables; engines and motors; other machinery	Some biomedical products	Other toys	
Process automation using micro-electronic technology	Some electronic products	Machining (batch and mass) especially in vehicles; consumer durables, and machinery; printing and publishing	Continuous-flow processes already partly automated; e.g. chemicals metals petroleum gas and electricity	Clothing; textiles; food; assembly	Building materials; furniture; mining and quarries	Agriculture; hotels and restaurants; construction personal services
Information systems and data processing	Specific government, business, and professional systems involving heavy data storage and processing in large organizations	Financial services; communication systems; office systems and equipment without total electronic systems; design	Transport wholesale distribution; public administration; large retailers	Retail distribution; all-electronic office systems; electronic funds transfer	Domestic households; professional services	Agriculture; hotel and restaurants; construction; personal services

[a] Ranging from less than ten years (rapid) to more than thirty years (slow) for the greater part of production to be affected.
[b] Proportion of total product or process equipment cost.

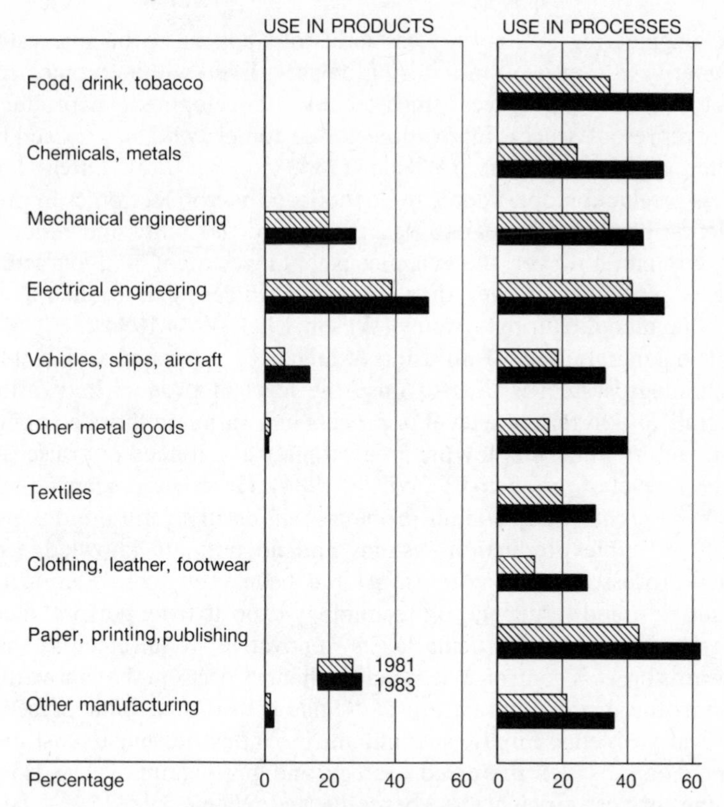

Figure 8.1. Use of Micro-electronics by Industry (Percentage of UK Factories)
Source: Northcott and Rogers 1984

the PSI survey that the most general application is that of programmable logic controllers for process control and testing devices for quality control. The results confirm the earlier work of Arnold and Senker (1982) that computer-aided design has been utilized more extensively in electrical and electronic than mechanical sectors; however, its use was also extensive in vehicles, with 39 per cent of sampled firms having CAD work-stations. Computer numerical control of machine tools is obviously concentrated in electrical and non-electrical engineering and vehicles, as are robotics and automatic handling devices (Northcott and Rogers 1984, 162–4).

Performance-maximizers may thus use micro-electronics in the development of new products, but the emphasis here is very much on the engineering sector: process industries appear to be interested primarily in improved monitoring devices. Even within engineering there may be differences, since technical development in products may represent process innovations for customers who may indeed be other engineering firms (Wilson 1984, 15). For both batch- and mass-production operations, the benefits of micro-electronics in production processes appear to be in increased flexibility and reduced set-up time; however, the evidence is that investment in manufacturing is piece-meal rather than involving wholesale introduction of flexible manufacturing systems (Wilson 1984; Voss 1984).

The general effect of diffusion of advances associated with a new technology system is thus to raise the level of product innovation overall, and to raise the level of process innovation in those competitive sectors which employ the greatest numbers: it need not raise the overall level of productivity (Wilson 1984). However, it seems likely that the greatest adjustment problems will occur in those industries with inflexible production systems and no intimate knowledge of microprocessor-based systems. As we have seen, cost-minimizing industries tend to depend on technology exports from performance-maximizing ones, particularly as innovative requirements shift towards process control. When major changes occur in the innovative sector, the incremental nature of change is disrupted. This presents several problems. Firstly, cost-minimizing industries must substitute innovation for their preferred strategy, and 'product innovation is the enemy of cost efficiency' (Abernathy and Wayne 1974, 121). Secondly, new processes must be adapted by managers who may be ignorant of the technology involved: case-studies show this to have been a problem in the UK (Winch (ed) 1983). Companies faced with these difficulties may avoid innovation by securing higher levels of performance from old equipment through cost-cutting exercises and by seeking to continue to compete on price (Utterback 1979, 56), or by setting up green-field sites involving new process or product variants: indeed, both approaches may operate together in multi-plant enterprises (Abernathy and Wayne 1974, 121; Abernathy and Townsend 1975, 391).

However, new technology may offer substantial benefits to successful innovators in mature markets, and these benefits hinge upon the idiosyncratic nature of micro-electronics. The essence of the Abernathy–Utterback approach is that maturity requires a trade-off between flexibility and efficiency. Many examples are quoted but the *locus classicus* is Abernathy and Wayne's discussion of the decline of Ford with the demise of the Model T. However, since microprocessor-based technology may reduce the costs of new product development and enhance manufacturing flexibility, it may facilitate a combination of cost efficiency and innovative behaviour by reducing the dedication of processes to particular products. According to Utterback and Abernathy, 'It may be that computer aided manufacturing will ultimately reduce some of the interdependence between product and process change' (1975, 646).

In manufacturing, the increased flexibility of robotic assembly affects economies of scale in both batch and mass production. Ayres and Miller have estimated that the potential for such cost reductions may permit price reductions of up to 30 per cent in US metalworking industries involved in batch production (1981, 58). In mass production, microprocessor-based process innovations may in effect alter minimum efficient scale and allow greater product differentiation and market coverage (Marsden *et al*., 1985). As Ayres and Miller note, the concept of flexible manufacturing is actually a development *beyond* the distinction between batch-manufacture and mass-production technology which combines the versatility of the former with the high operating rates and low unit costs of the latter (1981, 52). This has clear implications for the basis upon which corporations compete.

Abernathy *et al*. (1983) refer to the consequences of such 'new technology systems' (although they do not use the term) as 'de-maturity'. Once more the focus of analysis is the car industry, where the combination of technological change, oil-price rises, and the advent of Japanese competition is seen to have induced a renewed bout of process and product innovation in which the advantages of economies of scale are diluted by a renewed concern with product quality and performance. Crudely, 'de-maturity' can be seen as a movement *backwards* along the curves of Figure 4.1.

Similar considerations affect service industries. Although the framework of Chapter 4 related to manufacturing, in services, too, competition may occur on the basis of price or quality. Similarly, some services have a life cycle and are subsequently superseded by technical advance (Thomas 1978). Moreover, as Child has shown, similar concerns of cost, flexibility, and new service development influence the rate and type of adoption of micro-electronics (1984, 245–67). As Table 8.1 shows, microprocessors encourage the development of new services—for example, in maintenance and software applications—which are 'performance maximizing' as well as the development cost-minimizing process changes to do with information storage, processing, and retrieval in mature services.

The principal difference, which I shall discuss further below, is probably in the latter area. The scale of the savings in labour costs possible through wholesale adoption of electronic data processing is greater in many services than in manufacturing. Where one is concerned with the control of production processes, substantial investment in mechanical equipment is still necessary. However, in the white-collar areas which constitute much of service employment, cheap information-processing devices may supplant substantial amounts of labour. Chapter 2 has already indicated the strength of white-collar union responses to this threat, and the evidence has been of a more rapid rate of diffusion in services than in manufacturing (Brady and Liff 1983).

MATURE INDUSTRIES

The course of innovation is likely to be very different between firms in different sectors. Firms involved in new microprocessor-based product innovation are likely to be performance-maximizers: receivers of process innovation may be cost-minimizers. Figure 8.2 illustrates the two different innovative models appropriate to performance-maximizing and cost-minimizing firms respectively. In the former, in-house R&D proves to be the major innovative force. In the latter, although there are a variety of routes to innovation, the essential mechanism is the market.[1] In the one case, there may be familiarity with the innovation, predictability of impact, and the likelihood of business expansion. In the latter, managers may have to react quite rapidly to process changes with which they may be technically unfamiliar with no assurance that business expansion will result. The distinction is particularly important during phases of radical technological change. When an innovation has potential impact across a wide range of processes and products, many firms may find themselves to be process adopters where once they were product innovators. The relatively rapid advent of micro-electronics may thus force firms from the technology-driven model of change to the market-pulled one.

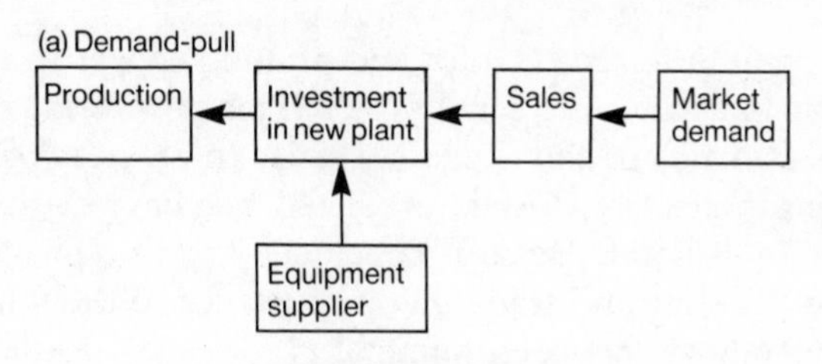

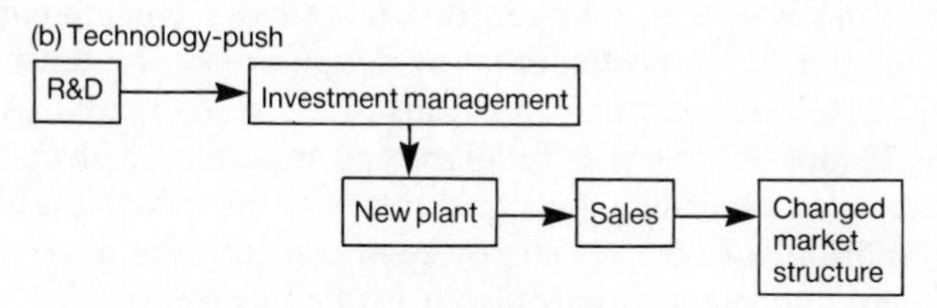

Figure 8.2. Demand-pull and Technology-push Models of Corporate Innovation

The two models have very different implications for trade union behaviour. On the most general level, it is worth noting that the two arms of the TUC's strategy discussed earlier—the promotion of R&D and the commitment to the negotiation of change—are probably directed at two different sectors. Whereas the concern with process innovators is with employment safeguards, the TUC's technology policy involves a concern with increased investment in electronics.

However, technology agreements are less likely to be achieved in mature industries involved in process innovation for a number of procedural and substantive reasons. The issue here is not trade union resistance to change in the face of employment loss. Employment contraction in UK manufacturing is a secular trend apparently unaffected by microprocessor technology (Wilson 1984). Indeed, contraction of employment and low output growth have even characterized the UK information-technology manufacturing industry which provides process innovations for mature industries (NEDO 1983). Moreover, survey evidence does not indicate that either job loss or resistance is widespread. In manufacturing, microprocessor-based applications have been associated with job losses of 0.6 per cent while resistance has been localized: textiles and printing appear to experience obstruction but only 7 per cent of the PSI manufacturing sample did so overall (Northcott and Rogers 1984). In services, such as banking and insurance, job loss is predicted but has yet to occur (Rajan 1984). In general, Bessant concludes that 'despite strong expectations to the contrary' trade union resistance to the introduction of new technology has been minimal (1982a, 23). Rather, the failure to reach agreement has to do with the speed and nature of change. *Ceteris paribus*, employers who react to change are likely to seek to move quickly in response to market competition; those who plan for it in advance may, by contrast, be able to accommodate trade unions' desires for involvement in the design process by consultation over in-house R&D efforts. This leads to the second point. As indicated in Figure 8.2, supplier relationships are often involved in process innovation. Recalling the arguments of Chapter 5, this is likely to lead to substantially greater employee uncertainty about the properties of new equipment. It may also frustrate involvement in design if suppliers will not co-operate in design modification.[2] Moreover, if managers are not technically competent, the scope for employee opportunism while the properties of new techniques are being learnt is substantial.

Overall, then, there may be particular problems to do with employee reactions to microprocessor-based process innovations in mature industries: such innovations are likely to be cost-reducing exercises, affecting large concentrations of employees who are probably members of trade unions, and there may be substantial scope for opportunism. Resistance to such changes may not be widespread, as indicated above; nevertheless, if acceptance is conditional, or co-operation is perfunctory, some of the benefits of change may be foregone.

However, a very different prognosis concerning the impact of microprocessors in mature industries exists. Sabel (1982), for example, suggests that, with technical advance, mass markets break down, production processes become flexible enough to deal in specialized innovative products, and 'Fordism' (machine pacing and task fragmentation) may give way to the emergence of high-trust organizations 'in which conception and execution are combined'. Managers may respond to market change by attempting

to use a combination of innovative technologies and organisational devices to increase the flexibility of production while holding to a minimum and sharply circumscribing discretion exercised at the workplace. (1982, 210–11).

Nevertheless, there remains the possibility that, in the terminology used here, technical change will lead to consummate co-operation, high-trust contracting, and the general enhancement of effort bargains.

Sabel's argument relates primarily to manufacturing, and indeed his ideas reflect those of other writers primarily concerned with the flexibility of machines rather than people. In essence, he suggests that the engineering argument concerning the reduction of differences between batch and mass production may be paralleled by a return to 'artisanal' forms of work organization.

There has been substantial research into the impact of microprocessor-based innovations, particularly in manufacturing. Overall, existing research stresses the piecemeal nature of change, the variety of managerial intentions behind it, and the importance of managerial choice; the lack of trade union resistance to change also emerges (Willman 1986).

Wilkinson (1983), for example, looked at several examples of change in batch engineering, in which managerial motives for change varied from simple enthusiasm for new technology to a concern to exercise more control over labour, and to the desire to launch a new product. He does not consider these motives in terms of the type of innovation being experienced by the firm. A more sophisticated discussion of managerial motives for change appears in Buchanan and Boddy (1983), but they are concerned to stress the complexity of motives within firms rather than variance between them. They distinguish strategic objectives behind change—which refer to the product-market—from operating objectives such as a concern with work in progress and from control objectives such as the desire to reduce operator error. In their cases, which included shipbuilding, tractor manufacture, and the food and chemical industries, they found that

Different management levels and functions had different expectations about technical change and the opportunities and threats that it presented. Senior Management and accountants concentrated on costs, return on investment, company image and competitiveness. Middle line management tended to concentrate on control of workflow. Supervisory management appeared to focus on reducing disruption of workflow and human frustrations. (1983, 242–3.)

Other writers have pointed to the distinction between motives for adoption of micro-electronics and cost justifications for it after the event. This has been particularly important in the introduction of CAD, where design flexibility and reduction of lead times were motives but justifications tended to focus on savings in drawing-office labour (Arnold and Senker, 1982, 5–7).

This work tends to be much stronger on descriptions of the process of managerial choice than in effective explanation of the forces constraining decisions. Wilkinson, for example, emphasizes the political nature of innovation decisions but rather tautologically accounts for variable outcomes (in terms of work organization) of (technically) similar changes in terms of managerial will or unorganized worker resistance (1983, 80–6). Similarly, Winch (1983) stresses the importance of 'innovation champions' without discussing the circumstances in which they might succeed or fail.

Work which does consider product-markets or competitive conditions does tend to find them significant. This is true of Martin's work on Fleet Street discussed in Chapter 6 and of other work discussing the car industry (Marsden *et al.* 1985) and British Telecom (Batstone *et al.* 1984). Sorge *et al.* (1983) found in their comparison of the adoption of computer numerically controlled machine tools in the UK and West Germany that batch size was an important influence on work organization: much as the approach of Chapter 4 would suggest, increasing batch size and increasing division of labour on programming tasks were closely related. However, none of these works takes a systematic or evolutionary view of product and process innovation.

This discussion of the impact of new technology on mature industries thus raises several very broad questions about the evolution of effort bargains where technical change occurs. In Chapters 9 and 10 I shall attempt to shed some light on these issues by looking in detail at the evolution of effort bargains in two companies experiencing microprocessor-based innovation. These cases are, in turn, drawn from two mature industries—cars and banking—in which investment in such changes has been substantial. The former is concerned with manual employees in a manufacturing industry where trade-unionism has traditionally been strong and resistance to change well documented. The latter deals with white-collar employment in a service industry where innovation has occurred without resistance over a long period of time.

CONCLUSION

Whereas the previous two chapters were concerned with examples of trade union resistance to change before the advent of micro-processors, the two which follow will focus on process innovations which have their basis in that technology. There are several reasons for this. The first has to do with the contention that there is something qualitatively different about these sorts of technical changes which implies an impact upon labour relations very different from that of previous generations of change. Since I have argued that the strategic intentions of companies are more important than the technology itself, this argument must be directly addressed.

The second point has to do with the economic conditions of the innovating industries. Although there were sectoral differences, in general the economic background to the forms of resistance documented above was relatively buoyant. By contrast, the early 1980s were a period of recession. Trade union resistance to new technology was thus a priori less likely. The terms of change are nevertheless very important, since the conflicts over change documented above arose primarily because of the inability of contractual forms to accommodate changing economic and technical conditions. Contractual arrangements established during innovation in a recession must be assessed in terms of their likely desirability and longevity if economic circumstances change.

In this chapter I have simply tried to show how this new technology fits into the general classification provided above. In these terms, both of the cases I shall discuss below are examples of 'receiving' firms to which process innovations, originating in other industries as new products, were diffused. They were both unionized and both, at the time of study, were concerned simultaneously with the control of costs and the development of new products and services: information technology had a role in both.

NOTES

1. The suggestion here is that 'technology-push' and 'market-pull' models of innovation are not mutually exclusive, even at the level of the firm.
2. This may be a particular problem in office automation or computer-aided design where standardized systems are marketed, particularly if the vendor is much larger than the purchaser.

9

Innovation and Efficiency in the Car Industry: Austin Rover Group

INTRODUCTION

THE approach I have adopted so far links process change, product strategy, and labour relations practice. In this and the following chapter I shall spell out these links within particular firms in order to provide detailed information on the mechanisms involved. Austin Rover Group itself is of interest for two reasons. Firstly, the ancestors of Austin Rover feature strongly in the analysis of spot contracting in the car industry in Chapter 7. Secondly, the company itself is a particularly interesting case, having accelerated through radical process, product, and organizational change in the 1980s: from operating a spot-contracting system with relatively stable technology in the early 1970s, it was trying to develop a labour relations strategy appropriate for one of the most advanced manufacturing systems in the UK in the early 1980s.

The chapter thus begins by describing the most recent technological changes in the UK car industry in general and Austin Rover in particular: it picks this story up where Chapter 7 finished. These changes are linked in turn to product and labour relations strategy, and the consequences for effort bargains, work organization, and collective bargaining are described. Despite substantial change to contractual arrangements, a number of tensions remain which relate to the issues raised by Sabel. The chapter thus closes by reconsidering the comparison between Austin Rover and Ford which was the concern of Chapter 7.

INFORMATION TECHNOLOGY IN THE CAR INDUSTRY

The manufacturing process described in Chapter 7 has changed substantially under the impact of new technology. In the UK, these changes occurred in the early 1980s, although overseas rivals had tended to innovate earlier. The stimuli for change have been variously identified as the changed market conditions in the aftermath of the large oil-price increases of the 1970s, awareness of the cost-cutting potential of micro-electronic advances under conditions of slow growth, the requirement to respond to legislative standards on safety and emission, particularly in the USA, and the emergence of the Japanese car industry to the front rank of world producers (Jones 1983, 14–16; OECD 1983, 53–5).

Product innovations may be classified as those which increase the power gained from a given amount of fuel, those which decrease body weight and air resistance (OECD 1983, 56), and those which increase the quality of the product. The first set involves changes to petrol engines, such as improved compression ratios and combustion, as well as the development of diesel engines and variable-speed transmissions. The second set includes change to the design and aerodynamics of the car body and the increased use of plastics and electronics, the latter both in instrumentation and ignition control.[1]

These changes obviously have very different sorts of implications for process change. Improvements to engine compression could be effected by machining on existing transfer lines, while the development of in-car electronics is likely to result in a higher percentage of bought-in components rather than process change within car factories. However, diesel-engine manufacture or changes to the sorts of gearboxes produced require substantial retooling of existing capacity, and developments in body construction have in practice generated the greatest process changes of the last decade. In general terms, though, the link between product and process change is close, whether the concern is with quality or cost minimization. For example, improved drag coefficients, a larger car body, and the absence of leaks are all much easier to achieve with more reliable automated welding, while labour and material costs can be reduced by improved product design.

Recent process innovations affect all the stages depicted in Figure 7.1. The stamping area has been subjected to incremental change, with presses becoming larger and robotic or automated transfer machines creating integrated press lines. The body-building area has been subjected to the most substantial change with the introduction of automated multiwelders, spot-welding robots, automated transfer, and computerized quality monitoring. However, the impact of robotics has also been substantial in paint spraying. Automated transfer lines and numerically controlled machine tools have operated in engine and transmission machining for a generation; the new generation of equipment is faster, more flexible—in that one can handle different engine-blocks on the same line—and thus more economical in the use of capital. Final assembly and trim fitting remain labour intensive and relatively resistant to automation.[2]

Two further changes of a more general nature appear to be having a substantial impact. The use of computer-aided design economizes on design costs, including labour costs, develops design options which economize on downstream assembly time, and offers the option of developing CAD/CAM systems which reduce demand for skilled manpower. Secondly, the use of new technology to effect general improvements in process efficiency is important: uses include computerized production scheduling and stock control, perhaps extending to the integration of component suppliers, automated inspection, and self-diagnostic fault finding on process equipment.

Technological charge does not, therefore, equate simply with the use of robotics: in fact, this generation of technological change affects investment requirements, demand for labour, economies of scale, and the extent of product differentiation in ways which, in turn, systematically affect labour relations.

A more rapid rate of product innovation implies shorter model lifespans and increasing returns to successful introducers of new techniques. Flexibility in new processes implies lower retooling costs within the lifespan of the equipment and thus a long-term reduction in investment requirements. The trajectory of change implies that contractual relations might need rapidly to accommodate substantial change but that thereafter the product cycle of employment typical of the industry in the past might disappear. Sporadic retooling for new models in the past implied insecurity, a disruption of work relations, and bargaining opportunities: under changed technological conditions this may cease to be the case.

The most substantial single reduction in demand for labour has occurred in the body-assembly area, with the introduction of robotics. Currently, therefore, the remaining assembly tasks in power-train, trim, and final assembly are the most labour intensive. Because product and process innovation go hand in hand, the precise extent of manpower reductions is always elusive: new products tend to be designed to absorb less labour, and hence, even without process change, manpower reductions at constant output might occur; however, qualitatively, demand has also changed, with an increased requirement for skilled maintenance workers.

The third area of impact is on optimal size. It is argued by some that the flexibility inherent in new technology can affect economies of scale in the industry by reducing the capital costs of model change. Given the capacities of flexible manufacturing systems, capital costs may be spread over several models or indeed several generations of models. Although this generation of changes would not remove the need for a common base of standardized components within a model range, and although scale economies in certain areas such as final assembly need not change, there may be a tendency towards the production of a large range of units, each at lower volumes, on a given line (Altshuler *et al*. 1982).

This offers insulation against shifts in demand for a given product, assists a higher degree of capacity utilization, and, together with the use of computerized inventory controls, allows inventory reduction. Taken together, these changes imply that car plants will be smaller (in employment terms) and more capital intensive, that they may need to adapt quickly to product-market changes, and that to achieve this companies will need the active co-operation of a generally more skilled work-force. Co-operation is essential because higher capital intensity and faster throughput of materials make plants much more vulnerable to strike action, and such action will cause more damage in the short term. Similar considerations extend to the other aspects of process efficiency described in Figure 7.2.

The major car manufacturers in the UK experienced this round of technological changes in the period 1979–84. Effectively, these events only concern Ford and BL among the main producers: the massive job losses at Vauxhall and Talbot shown in Figure 7.7 indicate their switch from full-scale manufacture to assembly of kits from continental plants. The main focus of investment has been in body-assembly, paint, and stamping areas, reflecting the availability of new technology, but there has also been investment in new engine and transmission production which are far more expensive process items (Hartley 1981, 19). All four major assembly plants have now been retooled in this way, and in each case product and process innovation went hand in hand. However, the retooling exercises differ somewhat in terms of their flexibility.

The least flexible automated line is the oldest, that at Longbridge, were the BL Metro is built. It was based on similar older installations at Saab and Renault, and can cope with only one body shell. However, the line, which relies on dedicated multiwelders rather than flexible robotics, can produce at very high volumes.[3] Longbridge began Metro production in 1980. By 1982, BL were installing a more flexible manufacturing system at Cowley, for production of the Maestro and Montego. In 1984, sixteen model variants could be produced on the line, and the company claimed to be able to produce 'all current and proposed Maestro and Montego derivatives in virtually any sequence', to incorporate design changes during the product life cycle, and even to produce new models on the same equipment.[4] As well as sharing highly automated body-framing lines, the different models share computer-aided design, engineering, and production control facilities.[5]

Over approximately the same period, Ford have retooled Halewood, for the Escort, and Dagenham, for the Sierra. They have, in addition, invested heavily in automated engine-machining lines at Dagenham and Bridgend.[6] Once more, the later system is more flexible, the Sierra line having high levels of automation plus flexibility of manufacture. The Halewood line was, in 1984, perhaps the least automated of the four.[7]

These automated systems are potentially highly productive, but because of this they are extremely vulnerable to disruption. In recognition of this, both car firms have sought changes in industrial relations to reduce the possibility of expensive equipment lying idle owing to labour disputes, and to ensure that the quality of product remains high. Austin Rover, BL's volume car producer, has in particular embarked on 'the most intensive employee involvement programme ever mounted in the car industry' in order to secure such objectives.[8] In fact, this search forms part of a distinctive labour relations strategy which has emerged over a long period of time.

INNOVATION, PRODUCT RANGE, AND PRODUCTIVITY AT AUSTIN ROVER

The component elements of what became BL in 1975 had suffered historically from under-capitalization, low productivity, poor management, a high strike rate, and the inability to achieve substantial scale economies. The UK industry as a whole had been subject to critical analysis by a number of government or government-appointed bodies stressing the need for reforms to improve performance (CPRS 1975; House of Commons Expenditure Committee 1975; Ryder Committee 1975), and in fact the recovery plan was funded by public money.[9] Although some of this input went to fund the cost of the programme of contraction which reduced employment in BL by 45 per cent between 1977 and 1982, the greater part went towards the development of a new model range and investment in new equipment to produce it.

Severe problems had emerged in the company's product range by the mid-1970s. Figure 9.1 compares the Ford UK and BL model ranges in April 1975. Model proliferation in BL related directly to process proliferation, with assembly being carried out on at least six sites compared to Ford's two: scale economies were not being achieved. Moreover, many BL models were reasonably long in the tooth. Figure 9.2 shows that by 1977 the 'core' models such as the Mini, Allegro, and Marina were in the declining phase of the product cycle, although sales of the Marina were lifted somewhat by the 'Ital' face-lift. Despite having a large number of models, BL were not particularly good at new product development: Jones notes that the company introduced only five new models between 1968 and 1977, whereas Ford UK launched nine (1984, 2, 12).

However, given the process changes in the industry, the product-led recovery required substantial new investment and, as Table 9.1 shows, the period since 1977 has seen substantial increase in several measures of capital investment. Although *absolute* levels of investment at 1968 prices have not increased substantially over the period—the retooling efforts stand out as exceptional—capital investment per employee and per unit of output since 1978 is very much higher. The reduced-volume BL is a more capital-intensive operation.

Similar figures focusing on the Austin Rover Group are unfortunately unavailable, but the experience of the passenger-car sector reflects that of BL as a whole except that, since Austin Rover has probably received the greater part of the investment and experienced the larger percentage manpower reductions, there may be some understatement of change in Table 9.1. However, the strategy pursued by BL and Austin Rover Group cannot be seen simply in cost-

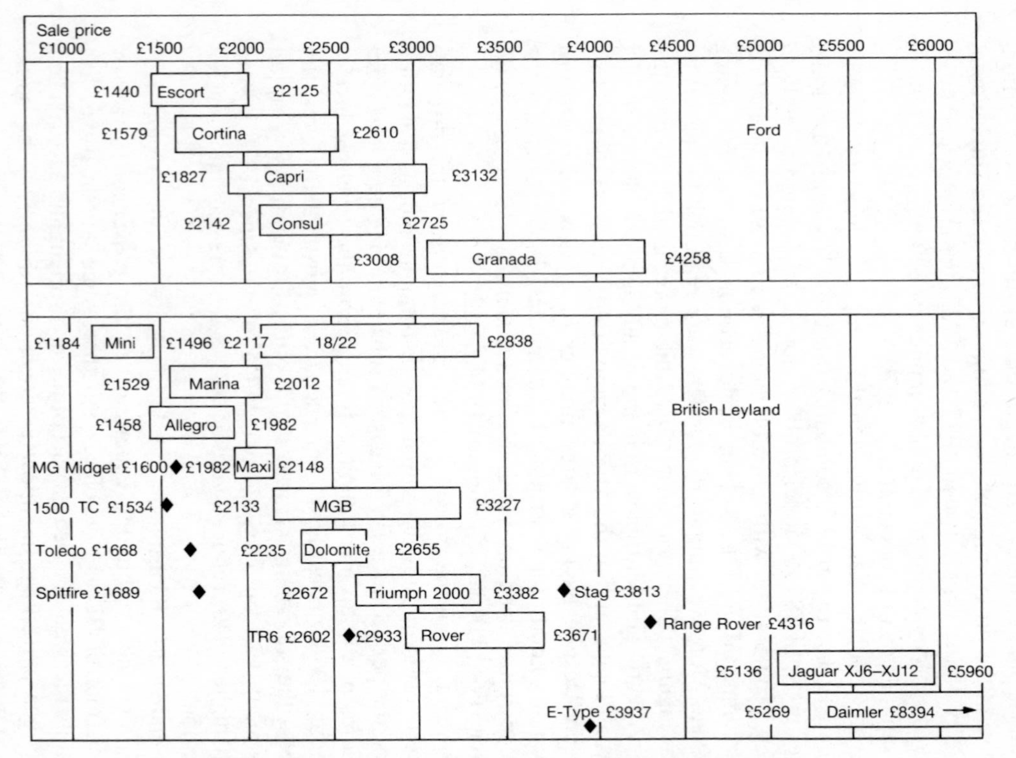

Figure 9.1. Ford UK and BL Model Ranges, April 1975
Source: BL

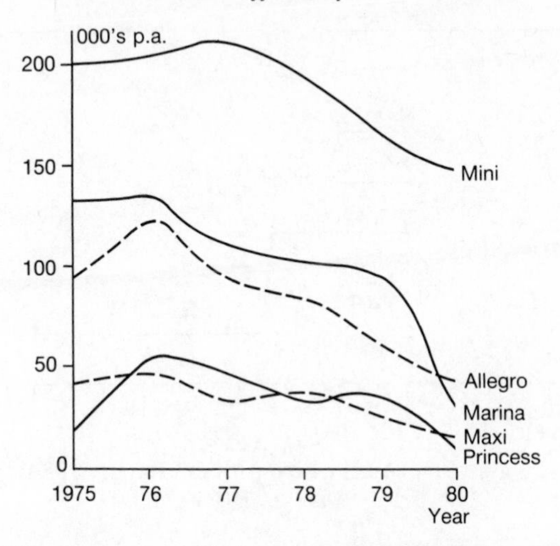

Figure 9.2. Production of Selected BL Model Ranges, 1975–1980
Source: Williams *et al.* 1983

Table 9.1. BL Performance, 1968–1983

Year	Vehicles sold (000)	Profit after tax[a] (£m.)	Employees (000)	Producivity (vehicles per employee)	Capital investment[b]	
					Per employee	Per unit of output
1968	1050	20.3	188	5.6	100.0	100.0
1969	1083	20.8	196	5.5	84.0	84.9
1970	984	2.3	200	4.9	101.2	114.4
1971	1057	18.4	194	5.4	72.0	73.5
1972	1127	21.1	191	5.9	56.9	53.7
1973	1161	27.9	204	5.7	72.6	71.1
1974	1020	(6.7)	208	4.9	106.1	120.4
1975	845	(63.2)	191	4.4	79.2	99.8
1976	981	46.5	183	5.4	87.7	91.1
1977	785	(5.0)	195	4.0	92.7	130.3
1978	797	(10.9)	192	4.2	136.4	183.2
1979	693	(118.5)	177	3.9	144.7	205.9
1980	587	(390.7)	157	3.7	151.6	226.1
1981	525	(339.2)	126	4.2	119.5	159.9
1982	519	(229.5)	108	4.8	146.9	149.4
1983	564	(74.3)	103	5.5	158.7	161.6

[a] Figures in parentheses indicate losses.
[b] Index, at 1968 prices (1968 = 100).

Source: Company Reports.

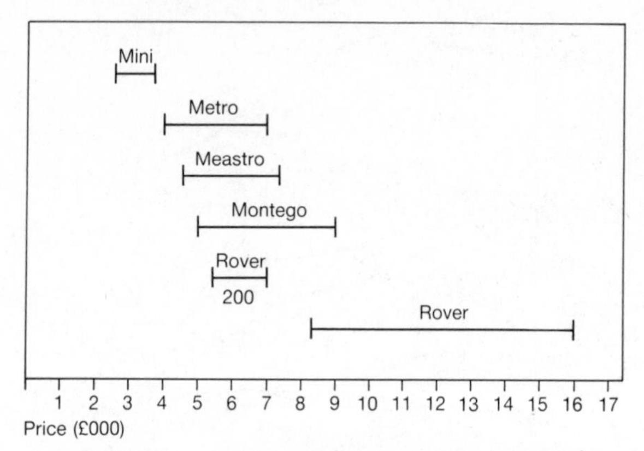

Figure 9.3. Austin Rover Model Range, 1984
Source: BL

minimizing or retrenchment terms, since the period of rationalization and process change also involved substantial product innovation. The model range in August 1984 is shown in Figure 9.3: since October 1980, Austin Rover has launched the greater part of an entirely new product range.

Over the period covered by Table 9.1, the company experienced three different labour relations 'regimes'. They are best characterized in shorthand terms by the names of the chief executives.

The Stokes Era, 1966–75

The period after the formation of BLMC was characterized by stagnation of output, investment, and productivity. New product launches—of the Maxi, Marina, and Allegro—were relatively unsuccessful and, following the oil-price rise of 1973, output and profitability fell markedly. Under-investment and the failure to rationalize production to achieve economies of scale during this period have been criticized by a number of observers (Williams *et al*. 1983; CPRS 1975).

As Table 9.1 shows, employment expanded during the period, but this followed directly from the failure of the labour relations strategy. The concern to devise a strategy appropriate to the rationalization of production within BL goes back as far as the original merger of the car companies into BLMC in 1968. In addition to proposing the transition from piece-work to measured daywork which itself was concerned with the reduction of shop-steward influence on the shop-floor, the strategy developed by 1970 included proposals for the reform of collective bargaining, a new range of consultative methods, and a concern to establish new forms of direct communication with employees, perhaps including the use of the ballots (Totsuka 1981; Ford 1972). However, the implementation of measured daywork throughout BL in 1971–2 left the right to negotiate manning levels with local trade union representatives: an essentially unproductive expansion of employment occurred after the change. Owen is able to show that, after 1973, unit costs at constant prices, which had been broadly stable since 1957, began to rise (1983, 62).

The Ryder Era, 1975–78

The deteriorating financial position of the company in 1974 caused sufficient concern for the government to appoint a Committee of Inquiry. On the basis of its report—the Ryder Report—the government effectively took BL into public ownership by purchasing the bulk of existing equity. The Report suggested reorganization of the company, continued government funding, and a shift of personnel. More fundamentally, it recognized the need for 'a massive programme to modernise plant and equipment at BL' (Ryder Committee 1975, 6) to launch a new rationalized product range. It also contained specific suggestions for the reform of payment systems and collective bargaining, and the introduction of industrial democracy. The concern with payment systems arose because of the failure of the move to measured daywork. The principal recommendation on collective bargaining was the specific suggestion that the number of bargaining units be reduced. In 1976, the thirty-four plants of BL Cars contained fifty-nine hourly paid bargaining units with renewal dates in nine different months of the year; BL as a whole contained 246 (Willman and Winch 1985, 70; Ryder Committee 1975). The pattern of competitive bargaining generated by this structure was seen to contribute substantially to the company's poor strike record. The recommendation on industrial democracy consisted of a proposal for a thoroughgoing participative structure, based on joint management committees at plant, divisional, and corporate levels. But the Report rejected trade union proposals for worker representation on the Board, and indeed the general purpose of the scheme was economic, i.e. 'to increase the effectiveness of the operation of Leyland Cars to the mutual benefit of all its employees'.[10]

The problem with the Ryder labour relations strategy stemmed broadly from the failure of the corporate strategy which embraced it. Between 1977 and 1979 the company operated a participation scheme and planned to centralize collective bargaining against a background of mounting losses and falling output: employment, too, began to fall steeply from 1978, and the company closed one of its Speke plants. This set up substantial tensions. There was a drastic loss of market share—from 25 per cent in 1977 to below 20 per cent at the start of 1979[11]—and the capital investment programme had yet to produce additional models and revenue to offset rationalization. The Ryder plan was essentially expansionist, the 1975 projection being output of 900 000 and a market share of 33 per cent by 1980: by 1978, it was well off-course.

The Edwardes Era and after, 1979–84

The end of the Ryder approach did not come with the arrival of Michael Edwardes in 1977, but rather with the announcement of his 'Recovery Plan' in 1979. The Edwardes plan called for a programme of capacity reduction involving the loss of 25 000 jobs and the closure or contraction of thirteen factories. The immediate stimuli for the change included a second energy crisis and an extremely strong pound which hit the company's exports, as well as the wider impact of the 1979 road-haulage and engineering disputes. However, the essence of the problem was the absence of new models and of rationalized productive capacity.

The Edwardes contraction plans were to precede a 'product-led recovery' which was projected to lead to an output of 950 000 units per year 'during the 1980s'.[12] Hence, as Table 9.1 shows, the period 1979–82 was one during which the capital intensity of operations increased markedly, even though employment and output both contracted. Subsequently, Austin Rover Group developed specific corporate objectives. These included a minimum 15 per cent return on assets and a 20 per cent UK market share: at 1982 UK sales volumes, this implied UK sales of 328 000.[13]

In addition, the company sought improved competitiveness and product quality, further simplification of the model range and the pattern of manufacture, 'the best technological strategies' available, and the development of collaborative relationships for major components with other producers (Evidence to Industry and Trade Committee, House of Commons, 1982).[14]

LABOUR PRODUCTIVITY AND THE PRODUCT-LED RECOVERY

Two broad strategic goals thus persisted throughout the Ryder and Edwardes periods: the launch of a new model range and the raising of labour productivity. The goal of labour relations strategy was focused on the latter. Throughout the 1970s, and particularly in the participation scheme, the company had assembled a range of information showing that a particularly severe form of competitive disadvantage lay in the system of labour organization: specifically, 'off standard hours' arising from disputes, restricted labour mobility, and demarcation were higher than in competitors' operations (Willman and Winch 1985; Marsden *et al*. 1985).

Success of the new product thus implied some reform of the system of labour utilization. This became clear at the planning stage of the first new model, the Metro, when the decision was made to invest heavily in new technology in order to 'maximise automation opportunities and reduce direct man hours per car to the lowest level possible. As early as 1976, the planners sought levels of efficiency which would require substantial changes in current working practices involving

> greater flexibility, multi-trade supervision, elimination of the stint system of working, eliminating late starts and early finishing, eliminating restrictive practices, the acceptance of some jobs currently done by indirect workers being carried out by direct workers (e.g. tool adjustments), the acceptance of full preventative maintenance and many others.[15]

A great deal of work-force co-operation in change was thus implied. Moreover, such co-operation achieved an even higher priority because of the concern with product *quality*. The Mini, for which the Metro was the planned replacement, had suffered from serious quality problems in paint, body-assembly, and trim areas, and the Metro's success was seen to be dependent upon their solution.[16]

The central problem for the labour relations strategy was how to achieve a highly productive, co-operative, and quality-conscious work-force. The Ryder strategy sought similar objectives through participative involvement in the running of the company. But trade union representatives could not be expected to co-operate in the contraction of the company and the loss of employment which the Edwardes plan proposed: on this, participation foundered. From the company's point of view, the escalating losses of 1979 required a rather drastic solution, particularly since there had been little success in the negotiation of the changed working arrangements mentioned above (Willman and Winch 1985, 110–29.). In practice, the company imposed rather than negotiated the changes, and secured higher productivity and quality through a combination of discipline and automation. Behind this approach lay both a philosophy and a strategy.

In December 1979, a document entitled 'Management in BL' circulated to managers at all levels in the company. The document stated that

> It is managers who have the responsibility for managing, leading and motivating employees and for communicating on company matters . . . shop stewards have the right to represent and to communicate trade union information to their members at the workplace but only within the rules and procedures jointly established.[17]

In a clear statement of the 'agitator' theory of conflict the document pointed to 'the small minority who would like to see BL fail', but committed the company to 'acknowledge and respect constitutional trade unionism wherever it operates within the company'.

The company's willingness to act upon this philosophy was soon displayed. A comprehensive package for changing working practices was circulating from autumn 1979 onwards and was eventually imposed in April 1980: it was circulated to employees as a 'Blue Newspaper'. The document was explicitly intended to 'supersede all other agreements, customs and practices', and it finally established centralized bargaining, a job-evaluated pay structure, and an incentive scheme. However, the key parts of the Blue Newspaper were designed to re-establish managerial control on the shop-floor.

The key item was the removal of 'mutuality'—the requirement that the company reach agreement with stewards about all changes before implementation. Prior to the Blue Newspaper, most plant procedures had contained the status quo clause of the 1976 EEF–CSEU agreement, which ran

> It is agreed that, in the event of any disagreement arising which cannot be disposed of, then whatever practice or agreement existed prior to the difference shall continue to operate pending a settlement, or until the agreed procedure has been exhausted.

The two most important areas of application were work allocation and working pace. In both of these areas, change was radical. The Blue Newspaper established the company's right to use all recognized industrial-engineering techniques at all times to establish work-speeds and manning levels: complaints were to be processed through a grievance procedure, but no status quo clause operated. For negotiated redeployment of labour, the document substituted the following:

> any employee may be called upon to work in any part of his employing plant and/or to carry out any grade or category of work within the limits of his abilities and experience, with training if necessary.

Whereas previously such changes would have required negotiations with stewards, after the Blue Newspaper they were to follow from management decisions.

The available evidence suggests that the new regime was established quite quickly, although not without resistance in some plants (Chell 1980; Willman and Winch 1985, 129–49). The company experienced a number of large disputes first at Longbridge and then at Cowley, over issues of work-pace and effort levels, associated with the launching of new models. As Table 9.2 listing the major stoppages in Austin Rover between 1980 and 1983 shows, similar issues tended to recur. Large disputes at Longbridge accounted for 20.5 per cent of working days lost in UK motor vehicles as a whole in 1980, and 25.7 per cent in 1981; those at Cowley accounted for 24 per cent in 1983. None of these strikes was overtly about new equipment, and in fact many of them originated on assembly work downstream from the automated body-assembly areas. However, a spate of non-wage strikes was associated with retooling exercises in both BL and Ford in

Table 9.2. Major Disputes at Austin Rover, 1980–1983

Date	Working days lost	Issue	Location
Apr. 1980	239 100	New working practices	Several plants
Apr. 1980	26 900	'Togging up' disputes	Longbridge
June 1980	26 800	Tea-break dispute	Longbridge
Nov. 1980	14 200	Seat-building 'riot'	Longbridge
Dec. 1980	21 500	Dismissal of 'riot' leaders	Longbridge
May 1981	44 500	Production targets	Longbridge
Nov. 1981	87 200	Pay dispute	All areas
Nov. 1981	148 300	Rest allowances	Longbridge
Dec. 1982	14 100	'Bad workmanship' dispute	Cowley
Mar. 1983	6 200	Job assignments	Cowley
Mar. 1983	125 700	Early-finishing dispute	Cowley

Source: DE Gazette.

1980–3 and, as I shall show below, the link with new technology was clear (Willman 1984).

The events of 1979–80 have been popularly characterized as Edwardes' victory over the trade unions. However, this is fundamentally misleading in two respects. In the first place, it was not wholly attributable to Edwardes' own approach. The company's pursuit of high productivity and high quality on new product lines through automation and the reform of working practices had been established as early as 1976 in the planning stages of the Metro project; the concern to remove mutuality goes back at least as far as the 1968 merger and may, as Chapter 7 argued, go back even further. The peculiar emphasis on management rights and shop-floor discipline pervading Austin Rover in the early 1980s may have owed more to a particular style of management than to the demands either of the product-market or the technical system of production, but continuous contracting through the exercise of mutuality could not survive the perceived competitive pressures. The style may have been that of Edwardes, but the strategy was determined by a long-standing requirement for high-volume production.

The second misconception is that the strategy was simply anti trade union or that some victory over unions was involved. In the first place, it is important to stress that the Blue Newspaper involved simultaneously an attack on shop-steward organization in BL and a reaffirmation of the company's commitment to what it termed 'constitutional' trade-unionism, identified by adherence to an institutionalized separation between in-plant consultation and corporate-level collective bargaining. It is perhaps best seen as an attempt to mould trade union organization to fit a radically restricted scope of collective bargaining. This is evident from the way in which procedural arrangements were re-established.

The new procedure agreement of April 1982 did not simply set out to write a constitutional basis for constitutional unionism. It affirmed the company's commitment to trade union membership for hourly rated employees and established check-off arrangements on a company-wide basis. Moreover, it specified the duties and rights of shop stewards in some detail: except when carrying out union duties, the steward was to 'in all other respects conform to the same conditions of employment as other employees'. Permission from the appropriate supervisor was required before a steward left his/her place of work; facilities such as telephone, time off with pay, and the possibility of full-time shop stewards were also included. However, the agreement also reintroduced the idea of consultation: after the confrontational phase of 1979–80, the company sought to secure work-force co-operation by manifesting its commitment to *a particular form* of unionism. The consultative arrangements were intended to secure the

> discussion of relevant information so that the views of the unions may be taken into account in the management decision making process.

This was seen as important because

> Employees and their union representatives have a major contribution to make to the success of the business. To make the most effective contribution, they need to know and understand the main facts concerning the performance and plans of the company.[18]

The logic was similar to that supporting the previous Ryder participation scheme; however, this was to be consultation rather than joint decision making. Committees were to be established at both corporate and plant levels. At both, they were to 'build upon established collective bargaining machinery' and to involve union representatives.

The two phases of the Edwardes plan thus had counterparts in the labour relations area. Contraction, and the laying of the groundwork seen as necessary for the new product launch, bred confrontation and the imposition of change, while the product-led recovery itself required a co-operative work-force producing high-quality output, for which a commitment to a more open style of labour management was seen as appropriate: co-operation was to succeed conflict. However, it is important to emphasize that the strategy of firm discipline was only one possible solution to the problems of product range and productivity identified by Ryder. It is difficult to avoid the conclusion that mutuality would have to disappear to secure targeted productivity levels, but the goal of a productive and co-operative work-force is conventionally assumed to be achieved by a rather greater reliance on consultation and negotiation than was in existence in 1979–80. It may well be that the rapid deterioration of the financial position in autumn 1979 effectively obliterated other options, but it needs to be stated that the labour relations strategy, while bearing a clear relation to corporate strategy since 1979, was not the only one available in the circumstances.

In fact, the phases of the strategy are not discrete. On the one hand, Austin Rover Group sought to maximize the efficiency of the new processes by labour relations reforms which appear to imitate those of Ford UK—management control and the eradication of local bargaining (Lewchuk 1984). On the other hand, the company sought to move into an era where such approaches were less relevant—namely the operation of a flexible manufacturing system employing a smaller more highly skilled work-force with considerable potential for disruption. This dichotomy introduced elements of instability into the pursuit of 'efficient' labour relations.

EFFICIENT LABOUR RELATIONS

In this section I shall attempt to assess the extent to which Austin Rover Group has achieved 'efficiency' in labour relations. In this assessment Gospel's (1983) division between work organization, industrial relations and effort bargains will be used.

Organization of work

Work-organization reforms of both a procedural and substantive nature accompanied the product-led recovery. It is useful to discuss them separately at the outset, since the distinction explains some of the remaining difficulties in work allocation and organization: while substantive changes required consummate co-operation from employees, the procedures for the organization of work imply perfunctory response.

The substantive changes implied by a move to more automated body-assembly techniques are exactly those predicted by Abernathy (1978): namely an increase in skilled indirect workers and a reduction in direct operators.[19] In the Longbridge plant, the percentage of maintenance employees in body assembly increased from 9 per cent on conventional lines to 26 per cent on the automated Metro line: the percentage of direct production workers fell from 75 to 59 per cent. Overall, labour savings were difficult to estimate, but on certain operations direct labour savings of up to 80 per cent were realized. Partly because of improved design and partly because of automation, direct labour man-hours per car for the Metro body were only 39 per cent of those for the Mini (Willman and Winch 1985, 60, 154, 156).

Although automation affected only one stage of the manufacturing process at this point, simultaneous improvements in process effi-

Table 9.3. Union Membership Changes at the Cowley Complex 1976–1983

Union	% Membership			
	1976[a]		1983[b]	
	Body	Assembly	Body	Assembly
TGWU	85.3	66.6	75.0	66.5
AUEW	13.1	31.1	21.6	32.4
EETPU	1.6	2.3	3.4	1.1
	100.0	100.0	100.0	100.0

[a] 1976 figures are for June for both plants.
[b] 1983 figures are for October for Cowley body plant, and December for Cowley assembly plant.

Source: TGWU.

ciency and design meant that rising output at the plant was associated with falling employment: whereas volume increased by 56 per cent between 1979 and 1983, hourly paid employment fell by 35 per cent. By 1984, Longbridge had one of the highest levels of productivity of any plant in Europe.[20] Less direct evidence is available for the Cowley complex. As Table 9.3 illustrates, AUEW and EETPU membership increased as a proportion of the total in Cowley body plant between 1976 and 1983; in the assembly plant, there was little change. Over the whole period, trade union membership approximated 100 per cent and there were few transfers of engagements. The main change appears to follow from the installation of automated body-assembly facilities in the interim: the downstream activities of the assembly plant were less affected. For the company as a whole, as Table 9.4 shows, the largest change over the retooling period has been the growth in the proportion of highly skilled workers, mostly at the expense of other indirect workers.[21] This rise is the more remarkable since, in the interests of process efficiency, the company reorganized the system of maintenance work to economize on labour on the more automated facilities. Maintenance workers patrolled the line in pairs—one mechanical and one electrical craftsman—in order to reduce response times to breakdown (Willman and Winch 1985).

Parallel efficiency gains were sought in the reorganization of direct workers into production teams. The central principle of the team concept is the organization of operations within a given production

Table 9.4. Grade Populations at BL Cars, 1980–1982

Grade	Population (%)		
	1980	1981	1982
1	14	16	19
2	16	16	14
3	51	50	50
4	17	16	15
5	2	2	2
	100	100	100

Note: Sample jobs within grades are:

1 Toolmaker; machine tool fitter; electrician
2 Paintsprayer; press setter; conveyor mechanic
3 HGV driver; direct operators in paint, trim, and body assembly
4 Fork-lift driver; storeman; fitter's labourer
5 Cleaner; sweeper; sanitary attendant

Source: Company documents.

zone around the foreman. All resources required for production within the zone come under the foreman, and responsibility for output lies with him. Under the foreman, a team structure, as depicted in Figure 9.4, operates on most of the new product lines. Within the teams, there is little demarcation between jobs and a certain amount of job rotation. Teams are responsible for rectification work and the performance of routine maintenance tasks. As well as being a unit of production, the team is also a unit of accountability and, more significantly, the zone within which the team operates is the employee's unit of attachment to the company:

> The concept is designed to ensure that problems within a zone are discussed and resolved within the zone. It is therefore inevitable that the zone will become more and more a focal point of communication to and involvement with employees.[22]

The logic of team-work is the pursuit of productive efficiency. The zone concept maximizes labour flexibility, helps prevent down-time of expensive capital equipment, and provides a framework for measuring labour input. However, the teams themselves lack autonomy: they are hierarchical in nature, built around the disciplinary functions of the foreman, and integrated into the assembly process.

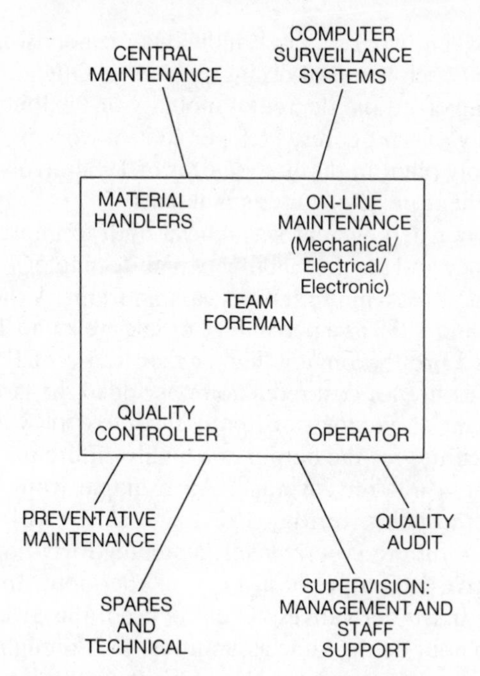

Figure 9.4. Structure of Production Teams at BL
Source: William and Winch 1985

Similar occupational shifts appear to have occurred elsewhere in the UK car industry, as well as in other European producers such as Fiat and Daimler–Benz (Marsden *et al*. 1985; Negrelli 1984; Jurgens *et al*. 1986). However, the changes to the procedures by which work was organized at Austin Rover are more distinctive.

The April 1980 changes had substituted managerial authority in the deployment and regulation of labour input for negotiation of changes through shop stewards. In doing this the company placed considerable burdens on the supervisory role, which the team organization increased even further. Problems arose because supervisors needed to secure higher work-rates in order to attain higher productivity, but the substantive reorganization of work laid considerable emphasis on the ability to evoke consummate co-operation.

These problems emerged periodically in the form of disputes. At least two types can be distinguished. Some, such as that at Cowley in December 1982 and that at Longbridge in 1984, explicitly involved problems between workers and individual supervisors over the enforcement of discipline or working standards. Others, particularly smaller ones, involved the degree of mobility or flexibility required: on the company's own figures, 91.5 per cent of working days lost at Cowley assembly plant in the first quarter of 1984 involved disputes over redeployment or new work assignments.[23]

Efficient labour deployment was crucial both to improvements in process efficiency and for the viability of high-technology production. In this area, labour-cost minimization was important. Although falling between 1979 and 1983 as a percentage of sales revenue, labour costs at BL remained proportionately higher than those of Ford.[24] However, to this pressure for cost reduction was added the largely incompatible constraint of gaining sufficiently flexible deployment to load high-output facilities for the output of a highly differentiated product to give maximum market coverage. An example from Longbridge may illustrate the point. In May 1977, a total of 9514 Minis were produced in *five* different derivatives: assuming forty-hour two-shift working, this gives an output of just over 29 per hour. In May 1983, 15 790 Metros in *ten* derivatives were produced: this gives an output of nearly 52 an hour on the same assumptions.[25] Although derivatives make little difference to labour deployment in the body-assembly area, in final assembly the pace of work and the complexity of man and materials management have increased considerably. Similar considerations apply to work at the Cowley plant, where assembly-line speeds on the Maestro were over three times those of the Rover line and twice those of the Acclaim line in 1983.

The high levels of conflict over redeployment—particularly in the period noted above prior to the Montego launch—resulted from the need substantially to revise a contractual requirement hitherto subject to joint regulation. Mutuality had implied something akin to an inside spot-contracting relationship wherein supervisors consulted stewards about manning and movement. Moreover, it implied a job property right—i.e. not only an expectation of confirmed employment but also the right to a *particular* job at a *particular* location (Turner *et al*. 1967, 333–9) and control over particular *items* of work.[26] The normal basis of attachment to a task or location which, under piece-work, could in effect be attachment to a particular effort bargain was seniority or custom and practice rather than operator efficiency or supervisory preference.

Teams do not exist in all Austin Rover plants, and objections to supervisory practice and redeployment are almost certainly not confined to team-work areas. However, the team concept does serve to illustrate, firstly, the organizational changes necessary to produce efficient working under changed technological and product-market conditions and, secondly, the problems encountered in the generation of consummate co-operation under conditions of change. As I shall show below, these problems have their roots in changes to individual effort bargains.

Industrial relations

The principal concern of Austin Rover Group in relations with trade unions was to secure the latter's observance of what was termed 'constitutional' activity. This term had a very definite meaning: it defined the 'voice' aspects of trade union behaviour. Even at the nadir of the company–union relationship in 1980 the company publicly stated,

> We believe firmly that the trade unions have a constructive contribution to make to the efficiency and success of BL cars, but can no longer suffer inefficiency and delay in the current round of negotiations.[27]

This contribution was seen to require corporate bargaining, a reliance on full-time officials, and the minimization of union influence on the production process. In turn, this implied close controls over the activities of shopstewards after 1980. One problem here was the difficulty of reconciling control over shop stewards with the desire to have stewards exercise controls over employees.

After the imposition of the Blue Newspaper, steward numbers and facilities were reduced at both Longbridge and Cowley, the two main manufacturing locations: in fact, events at both plants show remarkable parallels. The assault on shop-steward organization effectively began with the dismissal of Robinson at Longbridge in November 1979. Between 1980 and 1982 the number of stewards at the plant was halved from about 800 and 400 and the number of full-time stewards was reduced from eight to two. In April 1982 both the chairman and the secretary of the works committee were subject to disciplinary action because of their public disapproval of company policy (Willman and Winch 1985, 159–61). Similarly, at Cowley, withdrawal of steward facilities took place in 1981–2.[28] Five deputy senior stewards lost full-time status, and the number of stewards fell. In 1981 the convenor at the assembly plant was disciplined for calling an unauthorized meeting. In November 1982 a leading steward was dismissed. Overall, the number of stewards fell. At both plants, the primary mechanism was selective transfer of employees to new product lines where facilities were poor and supervision close. At Cowley assembly plant, by autumn 1983, 52 per cent of TGWU constituencies had no electoral nominees, and very few elections needed to be held: officials estimated after the election that over 30 per cent of employees were unrepresented by a steward.[29] The stewards' role had been further constrained by the rules on release from work in the 1982 procedure. However, steward organization survived in most plants, not least because the company's labour strategy required it: at no time do managers appear to have planned dealings with an unorganized work-force (Willman and Winch 1985, 149 f.; Chell 1980, 60 f.). However, the simultaneous experience of rapid change and a reduction in steward availability caused problems. At Cowley assembly plant in the first quarter of 1984, 22 per cent of disputes occurred over the unavailability of a steward. During one of the larger stoppages of the period, the company complained both that senior stewards appeared to have little control over sectional action and that they could not secure a return to work.[30] An open letter in May 1984 from the plant director called for an end to unconstitutional stoppages, remarking that

> Often against the wishes and recommendations of the senior stewards, small numbers of employees have stopped work and lost pay for themselves . . . your officials and your PLC are fully in agreement with the need to abide by . . . procedures.[31]

Several of these stoppages related to problems generated by the structure of 'constitutional' union activity. The central element in the company's approach to the latter since 1980 has been to bargain on as few occasions as possible. The bonus scheme was non-negotiable and the company has subsequently operated two-year pay deals. Pay in the car industry generally has continued a long-term decline against national averages into the 1980s: between 1972 and 1982, average hourly earnings in the industry fell from 123.8 to 103.2 per cent of the average for all industries and services (Willman 1984).

Dissatisfaction with bonus earnings was the source of periodic conflict, particularly at Cowley assembly plant, and clearly a regressive spiral of conflict could easily be set up: discontent over low bonus led directly to strike activity, since no negotiation was possible, and this further lowered bonus levels and raised discontent.

Effort bargains

The effort bargain at Austin Rover for manual workers is, in fact, dominated by the bonus scheme. The scheme, which operates at plant level and is non-negotiable, was introduced as part of the Blue Newspaper package: in theory, it can account for up to £30 per week, which in 1983 was over 25 per cent of the top skilled basic rate. At its core there are two calculations. The first calculates the *current efficiency* of labour in a plant by the ratio

$$\frac{\text{actual hours clocked}}{\text{standard hours of work produced}}$$

The denominator in fact refers to a quantity of work rather than a time elapsed: it describes the volume produced per hour at standard effort, i.e. the volume assessed on a non-negotiable basis by industrial-engineering techniques. The current efficiency then becomes the denominator in the calculation of the *efficiency index*, as follows:

$$\text{efficiency index} = \frac{\text{bonus threshold target}}{\text{current efficiency}}$$

The bonus threshold target at the outset was based on 1977 plant performance, and it thus produced anomalies between plants. Subsequently individual plants transferred over to Audited Plant Status (APS) wherein the bonus threshold was established by independent central industrial engineering audit. When the efficiency index exceeded 1, bonus became payable on the basis of average performance over the previous four weeks.

The first and most obvious point to make is that there are two ways of increasing bonus earnings on this calculative basis: one is to increase output, the other is to cut manpower. Using data from Cowley body plant, Towse (1982) is able to show that, from the company's point of view, increasing output generates the greater gain: the revenue from additional vehicles sold generally far outweighs wage savings at constant output. In fact, the evidence is that employment at Cowley and sale of Cowley models rose during the first three years that the bonus scheme operated.[32] Secondly, since it is plant-wide and covers plants which are links in the production chain, the scheme leaves the individual's bonus earnings vulnerable to the actions of others: there are thus potential difficulties about individual effort positions. Thirdly, it commits the company only to payment for cars sold (since the calculations are based on standard hours *shipped*), and

Table 9.5. Bonus Earnings[a] at BL Cars, 1980–1983 (£ per week)

Period[b]	BL Cars average	Unipart	Jaguar	Longbridge	Cowley assembly	Cowley body
1980–1	7.17	5.31	3.47	11.01	6.87	6.59
1981–2	16.80	9.96	16.68	20.08	12.66	17.21
1982–3	20.80	10.95	24.12	22.13	18.82	20.18

[a] The figures exclude bonus consolidation.
[b] Years are from November to November, i.e. between pay reviews (for explanation, see text).

Source: Company documents.

thus it does nothing to shield individuals from product-market fluctuations. All of these things introduce the potential for instability either of employment or of earnings.

In practice, the scheme operates very differently between plants. Table 9.5 shows average bonus earnings for the three largest plants—Cowley body, Cowley assembly, and Longbridge—for the first three years of the scheme. In terms of the (then) available maximum, bonus earnings were relatively low in the first year, rising steadily to November 1983. Throughout, bonus earnings at Longbridge, where output was dominated by Metro production, were higher than the average. The lower bonus performance of the Cowley plants was probably connected to the later launch of new models: whereas the Metro was launched in October 1980, the Maestro was not launched until April 1983 and the Montego not until April 1984. The figures show also the sales success of Jaguar being transmitted through to earnings.

The table understates both increases in earnings and the implied efficiency improvements. The November 1982 annual agreement provided for the consolidation in both November 1982 and November 1983 of £3.75 into basic rates, and for the raising of the efficiency index threshold above which bonus was earned to 1.101 and then 1.151. Consolidation thus allows for earnings improvement, but a given bonus level becomes harder to achieve. It is this aspect which led a TGWU official to refer to the scheme as 'a greasy pole' while a BL manager referred to it as 'moving the goalposts back'.[33]

Employees' earnings remained heavily dependent upon plant efficiency in the short term. However, as Figure 9.5 shows, earnings variations were affected by several other factors. Prior to the move on

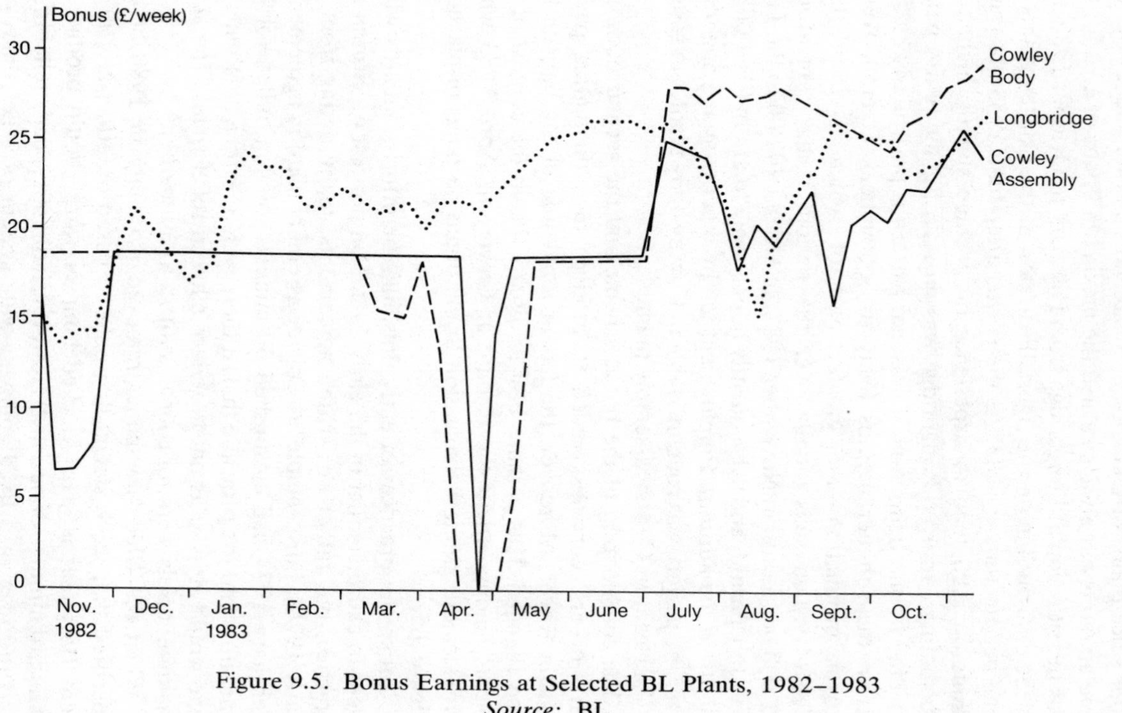

Figure 9.5. Bonus Earnings at Selected BL Plants, 1982–1983

Source: BL

to APS, efficiency variations did not tend to produce earnings variations since plants often overshot maximum bonus efficiency by variable amounts but always earned the maximum bonus of £22.50: this was the situation at Longbridge from February to October 1982 and, as the straight line on the figure illustrates, at the Cowley plants for long periods during 1983. As the period after June 1983 illustrates, plants on APS tend to experience on average slightly higher but fluctuating earnings. Longbridge was on APS for the entire period covered by the figure, and a seasonal pattern is particularly evident across the summer months (May to September). Overall, Austin Rover production—the greater part of which is from these plants—varied substantially over the period, rising from around 31 000 per month in November 1982, to around 40 000 in the January–July period, and subsequently falling away to around 35 000 per month in the autumn. Employment fell by 6 per cent at Longbridge over the period, was roughly stable in Cowley body plant, but rose by 17 per cent in Cowley assembly plant.

The combination of the bonus scheme and the secular decline in car workers' earnings tended to produce low, fluctuating pay at Austin Rover. Moreover, the labour input side of the bargain had also changed. It is perhaps best to discuss this in the context of the events leading up to the dispute at Cowley in April 1983 which provides the greatest fluctuation in the bonus earnings displayed in Figure 9.5.

Unlike the practices of early finishing, the effort standards which were to cause conflict in the early 1980s did not rest on custom and practice, but rather on written agreements, many dating from the early 1970s. For example, at Cowley assembly plant in agreement in February 1974 had established 50 minutes 'off line relief' for line operators and the principle that, if there were insufficient labour, the line would stop so as not to reduce allowed relief times. The actual manning levels were, of course, subject to mutuality.[34]

When the Blue Newspaper removed mutuality in 1980, it also specified that work standards would be based on effort at 100 per cent BSI: that is, each clocked hour of work should produce 60 standard minutes of output.[35] However, it did not specify relief times, and the idea of a fixed company-wide system of relief times only arrived with the implementation of the 39-hour week. In order to comply with the terms of the agreement that working-time reductions ensured 'that productivity is increased so that there are no increases in manufacturing costs as a result of the reduction in working hours.[36], the agreement reduced relaxation allowances to 40 minutes in the day. This prompted a lengthy dispute at Longbridge in November 1981 which was resolved by the establishment of a 46 minute allowance for direct workers, and 40 minutes for indirect (Willman and Winch 1985, 164–5); in addition, manning assignments were to operate at 101.5 per cent BSI on direct work.

These standards were implemented at Cowley in February 1982.[37] However, whereas at Longbridge most of the local allowances for 'togging up', 'washing up', or 'clocking on' had been removed either in the summer of 1980 or by the December 1981 settlement, at both Cowley plants it remained custom and practice—of about forty years' standing—to allow three minutes around start and finish times. In the run-up to the launch of the Maestro, the company sought to remove these allowances to secure maximum utilization of automated facilities: there followed a month-long dispute which was settled only with the establishment of an independent inquiry.

The issues involved were clear for both sides. For employees, over-standard working and the abolition of allowances increased effort input for a variable monetary output: it was one more assault on job rights. For the company, the revenue loss was crucial. Estimates of the precise losses vary, but the company suggested a loss of revenue on Cowley vehicles of £20 million per annum from early finishing: if one simply multiplies line speed by time lost from the practice, the annual unit loss is 2400 vehicles.[38] Since the company intended 'to achieve full use of its assets and investment and to produce more cars to sell', it argued the requirement for eight-hour continuous working involving *no* line stoppages.

The resolution of the dispute involved an inquiry into the assembly plant's poor industrial relations climate.[39] In evidence to the inquiry, the unions complained of poor housekeeping, excessive mobility of labour, poor safety provision, and a lack of trade union facilities. Management complained that stewards acted unconstitutionally, that trade union structure in the plant was inappropriate, and that there were too many stoppages. In their recommendations, the inquiry team mainly kept to the middle ground, finding some fault with both sides. The exception concerned the employees' complaint that they had to work too hard: the inquiry team responded that

> it must be recognised that there will be a continuing programme of efficiency improvements in the plant to secure its future viability. To this end there

must be a programme to inform employees at all levels of their involvement and contribution to the success of the plant.[40]

The company had already devoted some resources to this latter issue. As noted above, the 1982 procedure reinstated consultative arrangements. However, the concern with direct communication of company plans goes back further. Speaking of the period 1979–80 the (then) personnel director stressed the importance of the fact that 'every employee knew our side of the story at every stage'.[41] However, the zone system of working allowed a rather more systematic approach: monthly talk-ins, suggestion schemes, and the attempt to introduce quality circles all bear testimony to the concern to secure high-quality output through employee commitment. The company have argued that

> Employees have responded enthusiastically because they recognise that the Company has embarked upon a long term principled plan to improve the working lives of individual men and women.[42]

However, the communication of this commitment through efforts to introduce stability and security into the effort bargain was not pursued. There have been no moves towards the establishment of long-term earnings or employment security schemes: despite the development of internal labour-markets. It is difficult to resist the conclusion that, through the bonus and through the form of involvement without influence chosen by the company, the concern is to attach the employee to the success and life of the product rather than to provide longer-term security and earnings stability.

This receives some support from events at Cowley assembly plant after the resolution of the early-finishing dispute. On the Maestro line, the regime of slip relief and bell-to-bell work was established: on other lines, with products in the declining phase of the cycle, block relief, line stoppages, and short-time working were experienced. High capacity utilization of automated equipment remained the central issue.

The argument of this section is summarized in Table 9.6. In each policy area, there exist contradictory objectives and instruments: the threefold division into work organization, industrial relations, and effort bargains is for presentational convenience and it can be seen that the underlying problem is common to all three areas. On the one hand, the goals of productive efficiency can be served by machine

Table 9.6. Problems of Austin Rover Labour Strategy

Element of strategy	Objective	
	(Co-operation) Quality	(Control) Output
Industrial relations	Development of consultative arrangements	Removal of restrictive practices; management control of shop-floor
Work organization	Team-working; zone concept	Assertion of discipline; use of industrial engineering
Effort bargains	Involvement programme	Non-negotiable bonus

pacing of work, intensive supervision, output-related payments, and an attempt to keep trade union activity out of the plants. On the other, the requirements of high-quality output and production uninterrupted by disputes imply individual employee involvement, some form of team organization, and problem-solving activity by shop stewards. One might almost say that the two separate concepts of efficiency compete: the former is essentially an engineering conception which approaches labour relations with industrial-engineering techniques, whereas the latter acknowledges the motivational component of labour relations success.

IMITATION AND SUCCESS

Despite these inconsistencies, several indicators of performance hint at the success of the company's approach in its own terms. Between 1977 and 1983 strike losses fell, both in an absolute sense and relative to competitor firms (Willman 1984). Productivity rose and labour costs fell as a percentage of turnover; the value of work in progress fell. Figures on all of these items are presented in Table 9.7. The issue of quality of output is a little more difficult to deal with. Consumer surveys implied reliability problems with certain cars in the new model range,[43] but it is reasonable to suggest that overall quality standards had risen in the industry in the previous ten years: Austin Rover Group standards were themselves probably an improvement on past practice.

The comparative success of the strategy in terms of the removal of competitive disadvantage is more difficult to assess. The industry has

Table 9.7. Measures of Success for BL's Labour Strategy

Year	Strike losses (w.d.l. per employee)	Labour costs (% sales revenue)	Productivity (vehicle per employee)	Work in progress (% sales)
1977	17.3	27.0	4.0	20.9
1978	6.8	25.8	4.2	17.7
1979	10.5	26.7	3.9	18.8
1980	2.8	28.7	3.7	17.7
1981	3.0	25.2	4.2	15.1
1982	2.6	26.1	4.8	12.9
1983	1.8	24.5	5.5	10.3

Note: All figures refer to BL as a whole.

Source: BL.

changed so substantially that Austin Rover Group looks beyond Ford for comparators;[44] Ford, in turn, looks to Japan. Nevertheless, some similarities exist at the level of corporate strategy. Ford's 'After Japan' campaign illustrated the combination of concerns with automation, process efficiency, and quality outlined thus far.[45] In effect, the package was a mixture of financial targets and process changes. The targets were a minimum return on sales of 5 per cent, a minimum return on assets of 10 per cent, and break-even points for plants at 60 per cent of capacity: a minimum *European* market share of 14 per cent was set. However, along with this went a concern for the highest standards in performance and design of all volume manufacturers. These strategic volume/quality targets related to elements of work reorganization.

At plant level the key elements were:

(i) Reduction of inventories: an imitation of '*Kanban*' and a reform of scheduling and production control.
(ii) Increased automation and increased capacity utilization: this implied operator maintenance to reduce down time.
(iii) Higher proportions of production 'right first time': this involved self-inspection and quality concerns.
(iv) Manning flexibility: particularly on maintenance work.
(v) Reduced manning levels: including cutting relief time and introducing bell-to-bell working.
(vi) Reform of working practices: in particular, greater labour mobility and the introduction of a new disciplinary code.

Table 9.8. Product Differentiation[a] of UK Car Producers, 1975–1983

Year	Ford	BL	Talbot[b]	Vauxhall[c]
1975	12.0	5.1	5.4	3.7
1976	12.7	5.3	4.5	4.8
1977	12.2	5.2	4.7	3.1
1978	13.8	5.9	3.9	5.1
1979	13.8	6.6	5.7	4.4
1980	17.2	6.9	5.8	4.9
1981	19.4	6.7	5.8	5.4
1982	20.2	7.0	5.4	6.1
1983	16.8	6.3	5.6	7.8

[a] Figures are (no. of model derivatives)/(no. of basic models).
[b] Previously Chrysler UK.
[c] Includes UK-marketed Opel.
Source: SMMT.

These overall goals imply similar process-efficiency concerns to those documented above for Austin Rover, and a broadly similar involvement campaign: in practice, Ford have sought to introduce quality circles into their UK plants.

However, there remain two major differences. The first is that Ford UK remains a much more capital-intensive operation: despite the improvements in capital investment noted above, plant and equipment per head at Ford remained approximately twice that of BL throughout the retooling period 1979–83.[46] The second is that product differentiation of Ford is much greater than at BL: as Table 9.8 reveals, Ford get better market coverage through substantially superior performance in the production of derivatives. These features both seem to affect labour relations. Productivity per head at Ford, at 6.8 vehicles per man per year still exceeds that at BL, although the total of working days lost per head has been slightly higher since 1981.[47] Both labour costs and work in progress as a percentage of sales are much lower, although Ford UK sales are of course boosted by tied imports. As a consequence, although Ford shed labour across the retooling period while becoming a more capital-intensive operation, the pressure to relate labour cost to output in the short term through bonus is lessened both by greater capital intensity and by better market coverage. In this limited sense, then, capital intensity and labour contracting are more closely linked: process efficiency requires high capacity utilization, while the effort bargain is not based on the assumption that output variations will occur.

CONCLUSION

Presented with a spot-contracting system controlled by shop stewards, BL, and subsequently Austin Rover, sought to imitate Ford by establishing a complex comprehensive contract with a strong emphasis on managerial rights. The latter, as Chapter 7 stated, has probably been more efficient in the industry across the post-war period.

The first attempt to implement such a strategy failed. By the time the second attempt had succeeded, in the early 1980s, several further problems had arisen. Their implications for labour relations can be summarized by saying that the old-style Ford approach that BL had sought to imitate had itself become outdated in the search for consummate co-operation in the interests of cost and quality. Conceptions about process efficiency had, under changed technical and market conditions, come to rely in part upon employee involvement in quality. The strategy devised by Austin Rover Group did not wholly fit this need.

Nevertheless, the company's approach was comprehensive. Starting from perceived requirements for the organization of work, management designed effort bargains in the search for some mix of incentive and discipline to guarantee continuity of work and high productivity. This implied the attempt to eradicate a particular form of trade-unionism which seeks to control process operation while simultaneously seeking to encourage 'consummate co-operation'. The strategy thus supports a retention of company commitment to 100 per cent union membership, to formal grievance and consultation procedures, and to collective bargaining as the primary mechanism for the regulation of employment contracts. Austin Rover have sought to develop trade-unionism very much in the way that institutional economics would predict, but have not necessarily achieved the high-trust response required. Sabel's more optimistic view of the possibilities for work reorganization in mass-production industries receives little support.

One might argue that this is because the necessary conditions of information disclosure, democratic decision making, and security outlined in Chapter 5 did not apply, and that the case once more reflects the difficulty of moving from spot contracting to some form of relationship more suited to high-technology production. It is thus useful to turn at this point to an industry where several—though not all—of these necessary conditions do apply, and to look at the impact of microprocessor-based change in similar terms.

NOTES

1. This list has been compiled from Townsend *et al.* (1981), Jones (1981), Abernathy *et al.* (1983), and OECD (1983).
2. This may change. Austin Rover currently use robots to fit front and rear windscreens in final assembly; the OECD quote figures which suggest 12–13 per cent reductions in final-assembly jobs through the use of automated equipment in Europe and up to 50 per cent reductions in the USA, both by 1990 (1983, 100).
3. This paragraph relies on Willman and Winch (1985, Ch. 3).
4. 'The New Cowley'. Austin Rover Group, 25 Apr. 1984. The ability to face-lift or 'stretch' designs remedies a long-standing competitive disadvantage (Jones 1983, 21).
5. Sources: *Automotive Engineer*, May 1983, 77–80; Apr/May 1984, 53–7; *British Business*, 27 Apr. 1984, 756–8; *Financial Times*, 1 Mar. 1983.
6. *Automotive Engineer*, Dec. 1980, 63–6; *Ford Facts*, Apr. 193.
7. Willman and Winch (1985); *Engineer*, 23 Aug. 1982, 51–8; 7 Oct. 1982, 12. 'The Car Programme', document describing the Sierra launch, Ford Motor Company, probably Nov. 1982; *Engineer*, 23 Aug. 1982; *Automotive Engineer*, Oct. 1982, 39–42; observation, Dagenham engine plant, Apr. 1984.
8. 'The New Cowley', Austin Rover Group, 25 Apr. 1984.
9. By 1985 BL had received £2411 million. The greater part of this input was either government purchase of equity or loans later converted to equity.
10. 'Employee Participation in BL Cars', Company document, 1975.
11. Source: SMMT *Monthly Statistical Review.*
12. Quoted from 'Employee Communication from Company Chairman', 10 Sept. 1979.
13. Actual UK sales in 1982 were 281 600. The 1983 Company Accounts recorded Austin Rover UK sales at 346 600, exports at 91 000, and UK market share at 18 per cent.
14. These objectives were reaffirmed in 1983 (House of Commons) Industry and Trade Committee, *BL PLC*, London: HMSO, 1983).
15. This and the previous quote are from Company documents, 1976.
16. In these areas the percentage 'right first time' averaged 50 per cent on the Mini; targets for the Metro—achieved by 1981—were 85 per cent.
17. This and subsequent quotes are from 'Management in BL' (n.d., but circulated in Dec. 1979).

18. All quotes from BL Cars Procedure Agreement, Hourly Paid Employees, 1982. The consultative arrangements appear as Appendix D, of that agreement.
19. The latter change would, of course, follow simply from improvements to process efficiency.
20. Source: *The Engineer*, 9 Feb. 1984.
21. As Towse (1982) notes, over a similar period the ratio of *all* indirect to direct workers in the company actually fell, presumably because of the fall in semi-skilled indirect workers (i.e. grade 2).
22. Communication to employee representatives, Nov. 1983.
23. The other major losses were from disputes over the availability of stewards and support for the TUC Day of Action.
24. The comparison is as follows. Labour costs were 28.7 per cent of sales revenue at BL in 1980, falling to 24.5 per cent in 1983; as a percentage of total operating costs, the figures were 26.1 and 24.6 per cent. At Ford, labour costs were 23.8 per cent of revenue in 1980 and 20.7 per cent in 1983; as percentage of total operating costs, the figures were 24.7 and 21.1 per cent. However, in the latter case, tied imports form a proportion of sales.
25. Based on Longbridge plant management brief.
26. The 1969 mobility agreement of Cowley assembly plant contained the following: 'Where work is moved from section to section, line to line, or department to department, the labour employed on that work shall be afforded the opportunity to move with that work'.
27. Letter from Managing Director of the Cars Division to the general secretaries of all unions organizing hourly paid employees, 12 Mar. 1980.
28. Source: TGWU records. The issue of withdrawal of facilities at the assembly plant went to the Extended Plant Conference Stage of the Procedure, with the company offering no case.
29. TGWU documents, Nov. 1983.
30. Company document, 22 Feb. 1984.
31. Letter from plant director to all employees, 11 May 1984.
32. In 1981 (Aug) sales were 14 601 and hourly paid employment 6143. In Aug. 1983, sales were 29 626 and employment 9800.
33. Interviews, 6 Dec. 1983 and 25 June 1984.
34. Agreement, Cowley assembly plant, 27 Feb. 1974.
35. Blue Newspaper, Sect. 5.13. For a description of the effort-rating process, see Currie (1977, 57–231).
36. EEF–CSEU Agreement, Dec. 1980, Sect. 5.
37. The open letter to all hourly paid employees stated: 'To pay for the 39-hour week the Company requires track tied workers to have a fixed relaxation allowance of 46 minutes in each 8-hour day, a reduction of only 4 minutes, and have man assignments to operate at 101.5 BSI instead of 100 BSI' (12 Feb. 1981).

38. Letter from Chief Executive of Austin Rover Group to union leaders, 22 June 1983. Throughout, the company stated that early finishing was 'effectively' 38.5-hour working. The unit calculation assumes 48-week working.
39. The inquiry team consisted of two trade union and two company representatives, all from outside the plant.
40. Inquiry committee, *Report*, Recommendation 4.
41. Quoted from the *Guardian*, 10 July 1980.
42. Company document, Apr. 1984.
43. *The Times*, 5 Oct. 1984.
44. If BL's model was Ford, it was almost certainly Ford Germany rather than Ford UK. Commenting on the 1982 plan to the Trade and Industry Committee, Edwardes suggested that it 'is based on meeting the criteria of a highly successful continental competitor whom we identified in terms of productivity, manning approaches, work practices, fixed expense levels, direct and indirect (labour) levels' (House of Commons, Industry and Trade Committee, *BL Limited*. London: HMSO, 1982, 28).
45. Sources: *Ford Facts* 1983; Ford Motor Co. Presentation, Cranfield School of Management, 10 July 1984.
46. The ratios of Ford/BL plant and equipment (at cost) per head were: 1979, 2.14; 1980, 2.19; 1981, 2.17; 1982, 1.83; 1983, 2.01.
47. The average for BL for 1981–3 was 2.47 man-days lost per year; for Ford it was 3.73 (company figures). After retooling, Ford experienced demarcation disputes in its Halewood plant (1980 and 1982), the first major ones during the post-war period.

10

Information Technology and Collective Bargaining in Banking: TSB

INTRODUCTION

THE experience of computerization in banking has extended over such a period that one can speak of several generations of change within the industry. The major clearing banks first established a subcommittee to consider the automatic handling of clearing and computerized bookkeeping in 1955; since then, automation of clerical activity has proceeded more or less continuously (Wright 1978, 57). In the 1980s, a series of changes based upon microprocessor applications are being canvassed which threaten to change the character of the industry substantially.

In contrast to the industries discussed so far, experience of technical change in the banking industry[1] has occasioned little conflict. Over the periods discussed in Chapter 3, there have been no strikes over the introduction of new technology and a very low level of industrial action overall. Throughout the 1960s and most of the 1970s trade unions in the industry did not seek to negotiate terms for the introduction of new technology, and employment, productivity, and output growth coincided. However, the advent of the generation of changes associated with microprocessor technology has occasioned a change in trade union views: since 1979 the principal TUC-affiliated unions in the industry have sought unsuccessfully to establish new technology agreements, and the principal union, BIFU, has developed a policy of outright opposition to any changes which result in permanent job loss.

The industry thus illustrates a very different dynamic from those discussed so far: rather than offering examples of resistance to change and of conflict, it appears to have been a model of union acceptance. As one might expect from the argument so far, this history of acceptance is associated with economic circumstances and contractual arrangements very different from those already discussed. In the post-war period, employment in banking has grown steadily and contractual arrangements have generally offered substantial job security. Broadly speaking, the industry has moved from an authority relationship to a form of internal labour-market, though, latterly, microprocessor-based innovations appear to be encouraging the disintegration of the internal labour-market.

The industry is thus an example of the impact of radical technical change on internal labour-markets. I shall be seeking to show that the arguments of Chapter 5 are correct, and that, under circumstances of change, internal labour-markets cannot provide sufficient guarantees of employment and trade unions will thus pursue further safeguards. The structure of the chapter is as follows: first, the nature of technical and contractual change in the industry as a whole is described in relation to changing competitive conditions; secondly, the impact of such change on labour relations within a particular bank, TSB, is discussed in greater detail.

THE PATTERN OF CHANGE

A number of distinct changes can be identified in the process of bank computerization. In the first, large mainframe computers were employed to record customer accounts, print statements, and calculate interest or service charges. These installations mainly operated 'off line': i.e. data were sent on paper by road for processing at a computer centre and returned in the form of printed records for branch accounts. Most UK banks had adopted such systems by 1965 (Cockroft 1984; Lee 1973). The second major development was to put branches 'on line', allowing direct transmission of data between branch and the mainframe batch-processing computer. In most UK banks, this link was forged in the early 1970s. Once such a link was established, increases in the sophistication of VDU terminals and in the range of computerized bank services became possible. Terminals could cease to be 'dumb' and become 'interactive', the processing capacities of the mainframe could be extended by incorporating mini-computers at area or regional level into the system, and communication between branch and mainframe could be effected more rapidly: instead of information being batched at the end of the day, transmission of data could occur almost instantaneously (i.e. in 'real time') from counter terminals (IOB 1982; Revell 1983; BIFU 1983).

In addition, links *between* banks could be improved by the application of electronics to funds transfer and clearance. Automated clearance of non-cheque transactions has been the norm for some time, while high-value and international transfers in wholesale banking developed in the early 1980s with the establishment of the SWIFT and CHAPS systems: the former is an international message-switching system which automates international transactions; the latter performs roughly the same function within the City of London (Cockroft 1984).

However, from the retail (branch) banking viewpoint, the third generation of changes is associated with the use of microprocessor technology both to automate office work *within* branches and to direct customers away from branches. The first set of changes can be accomplished by the introduction of word processing and electronic mail; the second set follows from the installation of autoteller machines (ATMs) and cash dispensers, the more experimental usage of point-of-sale terminals away from branches, and 'home banking' based on Prestel or home computers (Cockroft 1984; BIFU 1983). The growth of ATMs in the early 1980s was the most spectacular of such changes, the numbers expanding from a few hundred at the beginning of the decade to over 7 000 in late 1984.[2]

Some of the other changes which would complete the set of computer links (Figure 10.1) are experimental rather than well established. However, their future development will continue the tendency for the branch to become merely one aspect of retail banking operations rather than the avenue through which all retail transactions are processed. Even without the development of home banking or point-of-sale terminals, the customer with both credit and ATM cards need not visit the branch routinely for cash. This is an important development, since to some extent routine cash or cheque transactions have been 'loss leaders' for the banks to attract customers to other services such as personal loans which *are* sold through branches (Meidan 1984).

Over the three generations of change, the objectives of innovation have shifted: whereas initially computerization was concerned with sales maximization, cost-minimizing concerns have become more salient. The three principal concerns of innovation in the 1960s were rapid business-volume expansion, consequential pressures on space and staff numbers, and the difficulties of expanding the latter in the tight labour-market conditions of the 1960s: hence the simultaneous expansion of output and employment during the first generation of change. Figure 10.2 presents the relevant data from 1971 onwards: in this instance employment and output data cover not just the banking

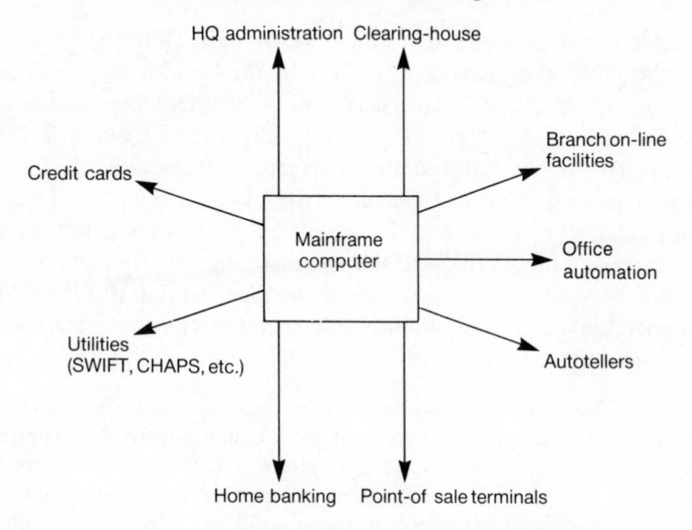

Figure 10.1. Computerized Links in Banking

industry but insurance and finance as well (SIC 81). It can be seen that employment has tended to expand rather less rapidly than output in recent years. The background to this needs to be explored.

During the 1970s and early 1980s, the purposes behind innovation shifted away from a concern merely to increase transaction volumes

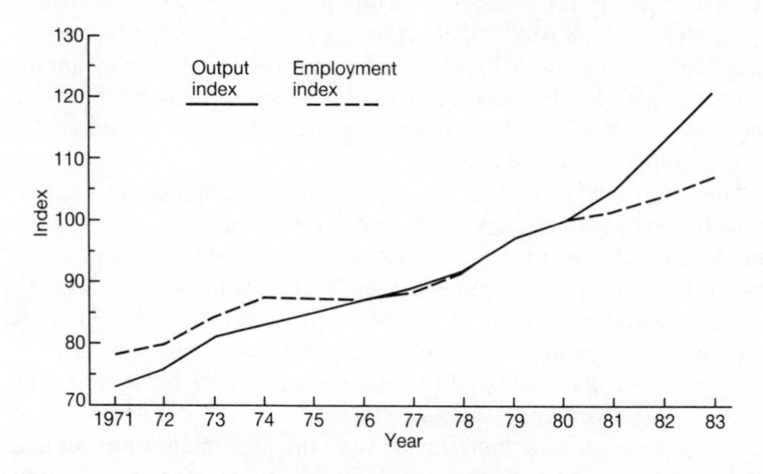

Figure 10.2. Output and Employment in Banking, Insurance, and Finance, 1971–1983 (1980 = 100)
Sources: DE *Gazette*; Income and Expenditure Blue Book

with a less than proportionate increase in staff. Between 1968 and 1979 the share of all UK deposits held by the London clearing banks fell from 33 to 28 per cent, while the share held by the building societies rose from 25 to 39 per cent (Shaw and Coulbeck 1983, 22–35). The banks entered the mortgage market after 1978, and regained some ground lost earlier to the building societies. Overall, the building societies led in the development of new products and services during the 1970s, and the trend amongst all financial institutions has been towards product diversification, so that different sorts of financial institutions compete in the same markets and the overall level of competition increases: since 1978,

> the rate of innovation has increased and interest rates have been changed more quickly. Rates have also been set more competitively. This contrasts strongly with the pre-1972 period when there was a limited range of options because of a low personal savings ratio, the importance of contractual savings, the bank control on rates and credit controls. (Shaw and Coulbeck 1983, 22, 16–27.)

At the same time, operating costs in banks within the OECD rose as a percentage of business volume: the UK clearers did not publish such information before 1980, but in the UK in both TSB and the building societies such costs rose steadily throughout the 1970s (Revell 1980, 167). Labour costs—a substantial proportion of operating costs in banks in the 1970s—rose rapidly and bank pay rose slightly faster than overall wage inflation in the UK in the latter half of the decade;[3] since most employees in the major banks operate within the branch network, the cost burden of sustaining high employment levels within the branch network rose, just as technical options for side-stepping it became available.

The role of information technology in banking—as in the car industry—is twofold. On the one hand, automation of money transmission attacks an item which accounts for 50–60 per cent of all operating costs (Price Commission 1978). Electronics can reduce the cost of cheque processing by 50 per cent (Cockroft 1984, 3) while various estimates show ATM cash withdrawals to cost substantially less than those made across the counter in a conventional way (Revell 1983, 66–7; Shaw and Coulbeck 1983, 62). Very broadly, the main cost reductions arise from reducing both the movement of paper and the amount of labour time required in a given transaction: the ultimate objective is 'the automatic initiation of all payments by the customer without the intervention of bank staff in the great majority of cases' (Revell 1983, 57).

However, as a cost-cutting exercise, the adoption of information technology is double-edged. In the first instance, there is a rise in fixed costs, a large portion of which consists of the amortization of development costs: in the short term, this puts further pressure on control of staff costs. The other element in inducing innovation is competition in the provision of new products and services: some products simply could not be offered in the absence of substantial new investment. For example, the monitoring necessary to provide interest-bearing current accounts, the provision of cash-flow accounts which automatically adjust funds, and—in the wholesale area—same-day settlement facilities depend upon electronics. For other services which have remained unchanged, the means of customer access has shifted to such an extent that a body of opinion within the industry talks in terms of a new 'product': for example, ATMs, which may appear to be a process innovation, introduce the new 'product', 24-hour seven-day banking (Cockroft 1984). The 1970s generally were a period of considerable innovation in the provision of services, the cost and quality of which could both be affected by the effective use of cheap processing power (Shaw and Coulbeck 1983, 42).

Information technology thus appears in the context of a dual concern to cut costs and to improve the range and quality of services. There is a direct substitution of capital equipment for labour in the operation of branches, but the role of branch staff in the marketing of newly developed services remains important in a market-place much more competitive than that characteristic of the first decade of computerization. New skills, and a form of consummate co-operation, thus need to be encouraged against a background of uncertainty. On the one hand, the career basis of the internal labour-market was fragmenting such that employers predicted that

> the rate of change of technology is likely to mean that the content of jobs will change much more than in the past . . . and it will not be possible for banks to maintain a policy of offering all staff long term career developmet. (Cowan 1982, 74.)

On the other hand, although the employment trend in the period of adoption of information technology remained positive, projections produced in the early 1980s were less optimistic (Table 10.1). Those

Table 10.1. Comparison of Banking Employment Forecasts

Forecaster	Publication date	Industrial coverage	Employment forecast
PACTEL	1980	Clearing banks	−10 per cent over 1980–90
Palmer	1980	Clearing banks	−5 to −20 per cent (under most likely scenario)
Kirchner and Hewlett	1983	Clearing banks	−5 to −10 per cent over 1983–90 (provisional forecast)
Shaw and Coulbeck	1983	Clearing banks	−12 per cent over 1981–90
Gaskin and Gaskin	1980	Scottish banks	−15 per cent over 1979–84
Robertson *et al.*	1982	Banking, advertising, business studies, and central offices	Minimum of +8 per cent over 1978–85, depending on state of economy
Rajan	1984	Banking	+5 per cent to 1987

which sought to predict developments in the longer term invariably saw staff reductions as inevitable. In this climate, trade unions in the industry sought seriously to negotiate over the effects of new technology for the first time.

THE EVOLUTION OF CONTRACTUAL ARRANGEMENTS

The contractual arrangements covering the first generation of change may be simply described as an authority relationship. Despite variations in size, all major banks operated a wage-for-age system with a standard grading scale covering those aged between seventeen and thirty-two. Above the latter age, increments were not automatic but depended upon individual performance and responsibilities (Morris 1984, 47). In addition, individual merit payments to staff on the grading scale could reward exceptional performance at management's discretion. Promotion from this clerical scale to a similar age-related managerial scale was possible, even likely, for many male employees, but unlikely for female clerical staff (Blackburn 1967, 71–2). Blackburn describes some of the conditions and requirements of bank employment in the 1960s.

> [the staff are] required to move wherever the bank chooses to send them and to work whatever hours are necessary to complete the day's work . . . the banks aim to treat their staff well, giving them good conditions of work as well as numerous fringe benefits such as special life insurance policies, low interest house purchase loans and a wide range of sporting facilities.
>
> One effect of these fringe benefits is to tie the clerk to employment in the bank. (1967, 77–8.)

Although there was some union membership, and collective bargaining in TSB from 1947 onwards, this form of contractual relationship persisted in the clearing banks until 1968, when national negotiating machinery was established. Indeed, elements of the relationship survived until 1971 and the establishment of a job-evaluated salary structure which was broadly similar across the major banks.

The events leading up to the establishment of collective bargaining have been described in some detail by Morris (1984, 49–114). Throughout the post-war period, the banks had to deal with continuous dissatisfaction over two aspects of pay: differentials between workers of different ages were reduced as recruitment pay rates were pushed up by relatively full employment, while overall there had been an absolute decline in real pay levels compared with the pre-war period. Pay discontent was exacerbated by the banks' tendency to increase pay by increasing bonuses—rather than pensionable basic rates—annually. Promotion opportunities had also decreased as the rate of growth of the banks fell during the 1950s. Because of this discontent, staff organizations emerged, and between 1953 and 1956 the major banks introduced arrangements for settling pay rates via negotiations with in-house staff associations. These arrangements culminated in binding arbitration, a device which was increasingly called for as overt disagreement between the sides emerged.

However, the clearing banks refused to recognize the main trade union in the industry, the National Union of Bank Employees (NUBE, now BIFU, the Banking, Insurance, and Finance Union), until the latter indicated the level of support it could command by calling the first strike experienced by the clearing banks in 1967. National negotiating machinery involving union and staff associations was established the following year, but the individualistic pay relationships remained.

The internal labour-market structure which provided the contractual background to the second generation of technical changes was established in 1971. Although there were minor differences between banks, the basis of the system was a four- or five-grade clerical structure in which rates were attached to measured jobs: minimum and maximum rates were established for each grade, and progression was to occur on the basis of ability rather than age.[4] Terminal and accounting-machine operators invariably were located in grade 1; cashiers and counter clerks, who at this stage were unlikely to perform jobs which were heavily affected by automation, were grade 2.[5] From the outset, staff in computer centres were generally on separate scales and rates of pay from clerical workers in branches, while a separate job-evaluation system was established for managerial staff. Salaries were governed by a grading appeals procedure in which a union nominee was involved.

The internal labour-market structure solved the staff problems of pay-differential compression and introduced equity into the effort bargain in that rewards corresponded systematically to some measure of labour input. The banks saw the system as conducive to career development and the encouragement of higher performance and as a means of reducing relatively high levels of wastage (Morris 1984, 232–3). There is now considerable evidence that inequality of promotional opportunity between men and women persisted (Egan 1982, Crompton and Jones 1984).

The persistance of such inequality is the key to understanding the change in contractual arrangements which accompanied the third generation of technological change. Figure 10.3 shows staff movements since 1971—the year job evaluation was established—broken down into full- and part-time employees. Since 1981, there has been a substantially greater increase in the numbers of the latter non-career staff in grades 1 and 2, almost all of whom are female. This is also the period during which the operation of cash dispensing, standing-order implementation, statement issues, and general clerical work has been affected by counter and back-office automation and the growth in number of ATMs.

Overall then, there is a remarkable correspondence between technological changes and changes to the contractual arrangements. In the first phase of automation, there was little branch impact and no need to change traditional paternalistic authority relations. In the second, the establishment of uniform branch routines based on computerized procedures was accompanied by the establishment of a job-evaluated, bureaucratic wage structure. In the third, where cost-minimizing pressures became important and staff economies became technologically possible, the internal labour-market structure has been eroded in ways which 'dual' labour-market theorists would predict (Piore 1971, 1979). However, these associations are tempting rather than empirically grounded: certainly those negotiating national job evaluation in the early 1970s did not feel that they were responding to technological change so much as establishing an equitable payment system for which there had been pressure for some time.

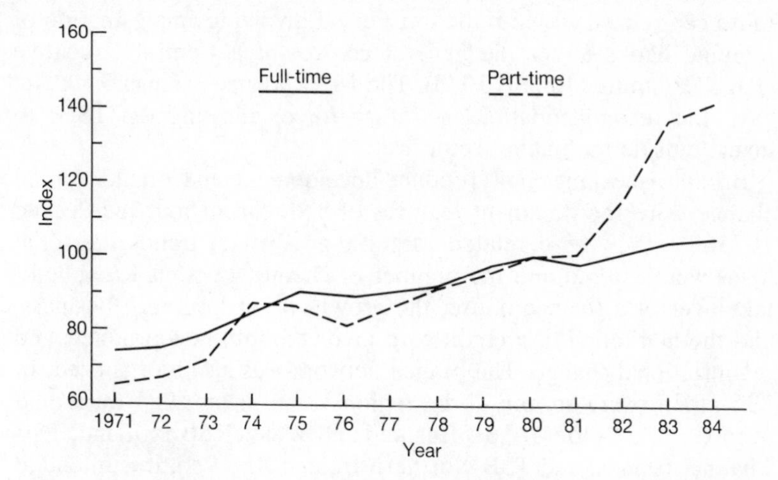

Figure 10.3. Full- and Part-time Staff in Banking, 1971–1984 (1980 = 100)
Source: DE *Gazette*

Nevertheless, in the third phase of change the union has become more aware of the problems of technical change. Since 1979, BIFU has unsuccessfully sought consultation and negotiation over technological change, proposing a model agreement based on that of the TUC. More recently—and perhaps more surprisingly—the umbrella organization of the remaining staff associations, the Clearing Bank Union, has expressed concern over the issue and proposed the establishment of consultative arrangements. BIFU's intention to resist non-negotiated change, fuelled by concern over recent redundancies (see below), is clear, although the union's capacity to do so may be in doubt. To understand these developments it is important to look more closely at the impact of change within a particular organization.

COMPETITION AND INNOVATION AT TSB

Background and systems

TSB, historically a collection of regional savings banks, changed substantially during the 1970s under the impetus of the recommendations of the Committee to Review National Savings (Page Committee). These recommendations were that TSB develop its range of products and services to provide the small saver and depositor with the full banking service appropriate to his or her needs and that a more centralized structure be introduced involving amalgamation of regional banks under the general control of a Central Executive (Page Committee Report 1973). The 1976 Trustee Savings Bank Act gave the recommendations statutory force, allowing the TSBs to invest outside the public sector.

Broadly speaking, new product development and organizational change were the dominant features of TSB throughout the decade 1975–85: they were related rather than distinct trends. Personal loans were offered and the number of cheque accounts expanded;[6] take-over of a finance house, the growth of an insurance business, and the marketing of a credit card involved both new products and organizational change. The branch network was also reorganized. In 1971 there were seventy-three regional banks: by 1984 these had been reduced to four—TSB England and Wales, TSB Scotland, TSB Channel Islands, and TSB Northern Ireland. In 1986, it is intended that a share flotation will consolidate TSB as a public limited company offering the full range of services to the retail sector.

Nevertheless, several features distinguished TSB from the 'Big Four' clearing banks. The first was simply size: in 1984 TSB had 25 000 staff (about 70 per cent of whom were employed in the 1500 branches), and was thus smaller than the Big Four. In addition, the savings bank tradition, and the terms of the 1976 Act, meant that the business remained focused on the personal sector despite attempts in the 1980s to develop a commercial portfolio. Thirdly, there remained a technical difference: TSB used an on-line real-time computer system which gave instant update of accounting information at the branch, whereas clearing banks used an on-line system which did not. There were both historical and business reasons for this which are discussed below. Finally, the pattern of labour relations in TSB had differed from that of the clearing banks throughout the post-war period: TSB had recognized NUBE since 1947, and since 1976 had operated a form of closed-shop agreement.[7] Terms and conditions of employment, including job-grading systems, tended to follow those of the clearing banks. In adopting computerized branch accounting systems, TSB was a pace-maker: indeed, savings banks in many countries have tended to develop more sophisticated systems before their adoption by clearing banks, in part because of their emphasis on the quality of personal service (Cockroft 1984; Turner 1980; Wright 1978). In terms of the technical stages discussed above, TSB omitted the first stage, moving straight to an on-line real-time system connecting branches to a central processor. This system was operative in many parts of the bank by 1971.

The basic features of the 1971 system, and its development since, may be briefly described.[8] At its core were four mainframe computers linked to simple terminals in branches: power resided in the computer, and the terminals, which were in appearance like large typewriters rather than VDUs, were 'dumb', in that all transactions required use of the telephone link to the main computer and the terminals did not respond to cashier enquiries. The system was 'on line' in that each branch was connected to a computer centre via a Post Office landline, and 'real time' in that the computer updated files and reported information to terminals immediately.

The choice of this system was based on several distinct considerations. The first was a concern to deal with a mounting volume of transactions without allowing quality of service to fall substantially. Before the system was installed, average transaction times were around five minutes, with cashiers facing long queues and experiencing customer frustration. After implementation, the average transaction time reduced to one minute (Wright 1978, 65, 68). Moreover, during the 1960s an increasing number of customers had sought to transact business at branches other than their own: some means of linking the branches was necessary. The second concern was cost. Staff numbers and costs had grown substantially in the 1960s (see Figure 10.4) and there was considerable pressure on space in smaller branches. The third main factor arose from the distinctive nature of TSB business in the 1970s and perhaps explains why the clearing banks did not adopt on-line real-time systems. The principal form of account in TSB was the passbook-based savings account which not only required that all transactions occurred across the counter (as

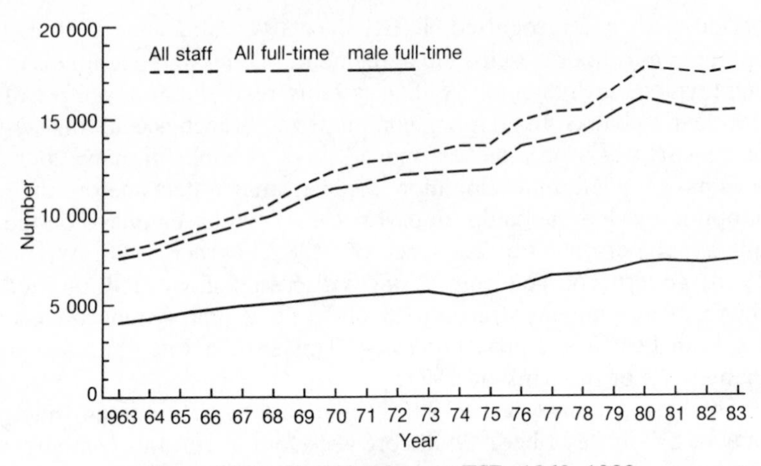

Figure 10.4. Staff Numbers at TSB, 1963–1983

Note: In the figure for all staff, each part-time counts as one employee. The figures exclude computer centres

Source: TSB

opposed to, say, cheque accounts) but also, prior to the 1976 Act, could not be overdrawn: real time allowed this security in the face of relatively higher volumes of counter transactions.

By 1978, problems of increased work-load required more central processing capacity, and the expansion of the range of products required more sophisticated terminals. In addition, there were cost problems which defined the parameters of the new system. The rising cost of the telephone link implied that some processing activity take place either at the branch or regionally; however, wholly distributed processing with micro- or mini-computers in branches was seen to involve too much expense on equipment. A movement away from on-line real-time processing would have brought TSB into line with other clearing banks and with business changes, but was seen to represent a reduction in the quality of service. The basic items in the new system established by 1980 are shown in Figure 10.5. A larger central processor links to a series of regional mini-computers which in turn link to intelligent reprogrammable VDUs with some processing capacity. In addition, the more sophisticated system supported an increasing number of ATMs. Simple cash dispensers had existed under the old system in one region, but the growth of the ATM network with facilities for investment, statement ordering, balances,

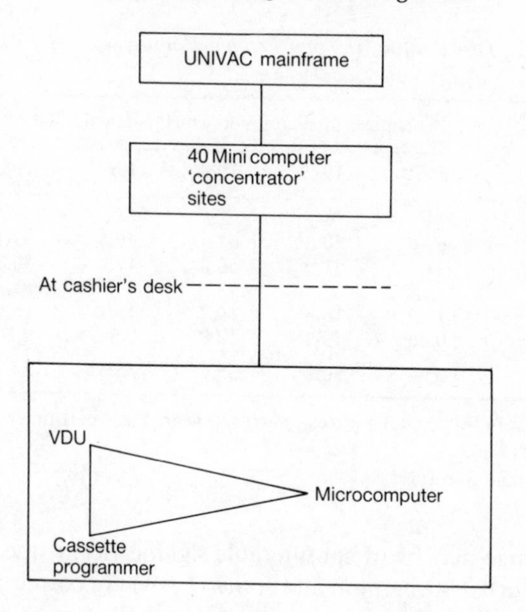

Figure 10.5. TSB On-line Real-time System, 1980

and other requests stems from the system change: the number of ATMs grew from ten in 1981 to 541 in 1984.

Throughout, the design of the system and the perceived nature of the market for services have been closely related. Currently, the four major strategic aims of the company are: to increase its share of the savings, personal credit, and insurance business; to improve the profitability of the money transmission business; to rationalize the branch network; and to improve the 'friendly and approachable' image.[9] As I shall show, the operation as well as the design of the system were tailored to these objectives.

Employment composition and the internal labour-market

Employment in TSB grew throughout the two phases of computerization (Figure 10.4). However, after 1963, the composition of the labour-force changed. The lowest percentage growth rates were in the number of male full-time staff: numbers of female staff rose by a proportionately greater amount. The largest growth rates were in the numbers of part-time staff, which increased with particular rapidity after the new computer system was installed.

Table 10.2. Distribution of Total Female Employees in TSB by Grade, 1978–1983

Grade	Percentage of each grade which is female					
	1978	1979	1980	1981	1982	1983
Clerical grades 1–3	69.6	70.3	69.6	69.5	69.1	69.2
Grades 4–5	31.3	51.9	50.8	47.2	50.2	46.0
Assistant branch manager	15.5	16.6	19.7	15.7	15.9	15.7
Branch manager	0.86	1.03	1.08	0.58	0.74	0.95
Overall %	55.4	56.3	56.4	55.3	54.8	54.5

Notes: Basis is full-time staff returns, new each year. Each part-timer is counted as one employee.

Source: TSB Staff Statistics.

This change may be of considerable significance. If the arguments of writers such as Crompton and Jones (1984) are correct, then these changes in the structure of employment over the period may represent a substantial growth in non-career staff, which in turn may indicate a declining number of career opportunities. Unfortunately data breaking down employment by grade, sex, and full- or part-time status are available only from 1978 to 1983; however, these data do lend some support to the proposition that the internal labour-market mechanisms did not operate universally. To the extent that this is the case, the accepted effort bargain within the industry breaks down.

Throughout the period, clerical part-time staff were employed exclusively on routine clerical and cashiering duties which fall within grades 1 and 2 of the full-time grading structure. Female full-time staff also tended to be employed for routine clerical or cashiering duties (Table 10.2) with little likelihood of promotion to assistant managerial or managerial level. For a growing percentage of staff, internal labour-market promotion mechanisms were either unavailable or inoperative. Moreover, for male clerical staff, conventional promotion opportunities within the branch network declined. Table 10.3 outlines some of the changes during the five-year period: the number of branches and of branch-manager jobs decreased, the number of lower-graded sub-branches and ATMs increased, while the number and percentage of computer-staff jobs showed no compensatory change. Although clerical turnover, particularly for

Table 10.3. Changes to the TSB Branch Network, 1978–1983

	1978	1979	1980	1981	1982	1983
No. of branches	1555	1554	1527	1520	1504	1488
No. of sub-branches	71	69	44	49	90	98
No. of ATMs[a]	—	—	4	10	56	255
Branch managers as % of male clerical[b]	34.5	35.5	32.0	32.0	31.0	30.6
Computer-centre staff as % of total staff	2.9	3.2	3.9	3.4	3.8	3.6
Female (grade 2) as % of all clerical	44.7	40.3	43.5	55.3	49.2	48.4

[a] England, Wales, and Scotland only.
[b] Includes assistant branch managers.

Source: TSB.

younger employees, remained, there was practically no wastage in managerial grades except that caused by retirement or redundancy: we shall discuss the latter below. In 1983, 89 per cent of assistant branch managers and 55 per cent of branch managers were under forty-five, in fact only 10 per cent of the branch managers were within ten years of retiring age. The bank's own projections suggested that the situation would deteriorate in the 1980s, with over 70 per cent of assistant branch managers being under thirty-five by 1985.[10]

As a response, TSB began to tier recruitment. Career and non-career recruitment literature was produced, the principal differences being that the former emphasized 'growing from grade to grade' and the acquisition of professional qualifications, whereas the latter argued that cashiering work was a 'worthwhile job'.[11] The planned recruitment mix was non-career (mainly school-leavers and married women returning to work), part-time staff, much smaller numbers of graduates on management trainee schemes, and *ad hoc* mid-career recruitment.

The growth in ports of entry indicates the disintegration of the internal labour-market. In the terms used here, these changes fundamentally alter the effort bargain. Internal labour-markets, it will be recalled, have incentive properties because of the prospect of advancement and because the idiosyncratic nature of exchange increases over time, so offering the prospect of high-trust relations. Where no advancement is offered, the bargain in truncated and the level of predictability for the employee is reduced. It is thus possible that a very different form of commitment may result. In order to understand the mechanisms involved here, it is necessary to look at the structure and flow of work within the branch.

Technological change and work organization

Overall in the TSB at the end of 1983, counter transactions accounted for 43 per cent of branch work-load and were the largest single item. Clearing accounted for 16 per cent and dealing with mail accounted for another 19 per cent; only 3 per cent of work-load was accounted for by lending.[12] Although there was some variation between branches and between regions, in all, counter work was, despite high levels of automation, the largest single work-load element.[13] Counter transactions were also the most heavily 'paced' elements of work. There were a number of reasons for this. First, the objective of the system designers was to reduce queue lengths and provide a better counter service: this translated into a concern to reduce transaction times for any given service. (Cheque withdrawals, for instance, under the newer system were work-studied at 1.29 minutes.) These times were set to keep customer waiting time down during peaks of activity, but it did not necessarily follow that slack periods would result in longer cycle times since counter manning was geared to analysis of customer arrival patterns: hence fewer terminals would operate in slack periods.

However, the costs of the system itself demanded continuous throughput of transactions to prevent expensive equipment lying idle. Hence there was pressure to put more and more services onto the computer system. This served to direct work away from the 'back office' as many procedures—for example, credit scoring—could be dealt with at the counter. However, since many of these newer services had much longer transaction times, this tended to conflict with the goal of reducing queues at the counter.

This problem was exacerbated by several staffing implications of computerization. The first was that the bottom 'general duties' grade was removed. Under the old manual system, clerks used to spend twelve months on back-office duties before progressing to cashiering duties. By 1983, new recruits typically spent a three-month period as 'supernumeraries' in the branch, during which there was a two-week off-the-job training course on the computer system. This often meant that the counter was staffed by relatively raw recruits with a relatively poor knowledge of the range of banking services.[14] In the same year, 68 per cent of all clerical staff in the branches were in the cashier grade 2. This concentration had arisen after the establishment of grade 1 (originally the grade above 'general duties') as a training grade in 1979. Work was then reorganized so that more complex transactions were re-routed to a more experienced clerk at a seperate counter station.

The desirability of ATMs followed from similar considerations. Transactions executed by customers on ATMs not only reduced pressure at the counter but also reduced back-office work such as cheque clearing. In addition to improving process efficiency, they also provided a new 'product' by initially circumventing restrictions on opening hours. However, the main criteria for ATM location in TSB in 1981–2 concerned the volume of transactions which would be processed through a given machine, and the volume of displaced counter transactions and hence staff savings:[15] large branches tended to get ATMs.

Moreover, ATMs continued to encourage the changes in the nature of employment discussed above. Table 10.4 contrasts staff changes in a sample of ninety-six branches in TSB England and Wales which acquired ATMs in 1982–3 with those experienced in the bank as a whole: while, overall, full-time staff numbers hardly changed, and part-time staff numbers rose by about 6 per cent, the ATM sample showed a fall in numbers of full-time staff and a huge percentage rise in part-timers. This effect was not continuous. Staff changes in these branches in 1983–4 were broadly in line with those for the branch network as a whole.[16] Nevertheless, there is some reason to believe that the sudden change was at least indirectly related to ATM

Table 10.4. Changing Staff Numbers in TSB, 1982–1983
Percentage change

	TSB overall[a]	ATM sample (n = 96)
Full-time	1.4	−3.5
Part-time[b]	5.8	50.9
Overall[c]	2.0	3.2

[a] England and Wales branch network.
[b] Counted as wholes.
[c] Each part-timer is counted as one employee.

Source: TSB.

Table 10.5. Work-load and Staffing Changes in TSB, ATM Sample (Correlation Analysis)

	ATMS	TIME	TRANS	VOL	TRANS/VOL	STAFFL	STAFFC	PT
ATMS	1.00	0.268[a]	0.596[a]	0.657[a]	0.206[b]	0.795[a]	0.005	0.271[a]
TIME		1.000	0.371[a]	0.265[a]	0.273[a]	0.304[a]	0.027	0.207[b]
TRANS			1.000	0.634[a]	0.701[a]	0.583[a]	0.063	0.098
VOL				1.000	−0.032	0.768[a]	0.034	0.224[a]
TRANS/VOL					1.000	0.101	0.013	−0.049
STAFFL						1.000	−0.100	0.295[a]
STAFFC							1.000	−0.005
PT								1.000
Variable means	1.27	3.06	170.7	51 081	0.180	16.79	0.35	0.774

Notes:
[a] Significant beyond 0.01.
[b] Significant beyond 0.05.

Variables:
ATMS: Number of ATMs operative in each branch in May 1982.
TIME: length of time, in months, of operation.
TRANS: volume of ATM transactions, May 1982.
VOL: volume of counter transactions, 1982.
TRANS/VOL: ATM transactions as a proportion of counter transactions.
STAFFL: Staff level, Nov. 1982 (each part-timer counts as 0.5 of an employee).
STAFFC: Staff change, 1982–3 (each part-timer counts as 0.5 of an employee).
PT: Change in numbers of part-time employees, 1982–3.

Source: TSB.

installation since, within the sample, changes were closely related both to branch size and to ATM usage.

Table 10.5 presents the results of a correlation analysis covering five measures of work-load and three of manpower and manpower change. The number of ATMs, their utilization, and the length of time they have operated are all significantly related and their relationship with the volume of counter transactions is strong. This reflects the tendency to site ATMs in large branches and the trend for ATM usage to increase over time. Transaction volume and staff numbers may be seen as two related measures of branch size.

The analysis provides little support for the view that the introduction of ATMs is associated with staff savings. The sample actually experienced staff growth of 3.2 per cent against that of 2.0 per cent for the group as a whole, and the change in staff numbers was not significantly related to any measure of counter or ATM work-load.[17] Since these measures relate only to a fraction of branch work-load, this is to be expected. In fact, although cash withdrawals continued to decrease as a proportion of work-load—on average for the sample by 9.1 per cent to May 1984—this change correlates neither with change in ATM usage nor with the two staff-change measures. Other elements of work-load were being introduced into branches.

However, changes in part-time staff were significantly related both to branch size and to two of the three measures of ATM usage. On this basis, it seems reasonable to conclude that the growth of ATM usage may accelerate the growth of part-time employment. Both the location of ATMs and thus the nature of the sample are important here. Average branch employment for the sample was 16.79 in 1982 and 17.14 in 1983. The figures for TSB England and Wales as a whole were 10.6 and 10.9 respectively. Reflecting the ATM installation policy, large branches tended to get ATMs whereas small ones did not. However, such larger branches tended, prior to ATM installation, to have relatively few part-timers. In the ATM sample there were seven full-timers for every part-time employee before installation, and 4.5 afterwards; in the bank as a whole, the ratio stood at approximately 4.5 : 1 in 1982 and 4.3 : 1 in 1983. Installation thus brought staff composition in large branches into line with that of the bank as a whole.

Two possible explanations for the large growth in part-timers emerge. One relies on technical change, but the other merely on improvements in process efficiency through work measurement. In

1983, the work-measurement scheme—the 'Branch Establishment Scheme'—was improved and the amount of work covered was increased: it could be that the impact of improved work measurement encouraged a move to part-time employment in larger branches, although this is not the view of the bank itself. Certainly the staff impact was short lived. Changes in part-time and full-time staff in the following year, 1983–4, are actually *negatively* related to those in 1982–3.[18] However, there is some doubt about the extent to which the work-measurement scheme could actually dictate staff levels in branches: for this sample, there was actually a 30.9 per cent positive discrepancy between actual and work-measured staff levels in 1984. The data is by no means conclusive, but because of the distinctive trends within the ATM sample, and because the link between staff changes and ATM growth can be demonstrated more closely than that with the Branch Establishment Scheme, it seems reasonable to suggest that ATMs account for much of the observed part-time growth.

The links between ATM operation and the growth in part-time staff appears to be as follows. ATMs may displace transactions from the counter in at least three ways: customers may defer cash withdrawals until after bank closing; they may use ATMs instead of the counter at their branch during opening hours; or they may go to a branch other than their own in order to use an ATM. Evidence from computer returns indicates that ATM usage of the second type is substantial throughout the working day and that a relatively constant level of transactions is displaced, enabling part-time cashiers simply to cope with peak counter loading.

Perhaps one reason why full-time staff numbers have not fallen substantially is that the ATM requires additional work itself,—in the marketing of cards, in maintenance, and in response to leaflet requests. Given the tendency of ATMs to generate transactions rather than just displace them (Willman and Cowan 1984), industrial engineers tended to view ATMs as product as well as process innovations.[19] However, it was perhaps much more important that the transactions displaced—cash transactions or cheque accounts—were not the largest work-load elements in any of the ninety-six branches: even at the counter, deposit account and investment account business were important. Moreover, the product range was growing as the ATMs were introduced.

This pattern of change raises several theoretical issues. The general picture of automation, pressures on efficiency, and routinization of procedures which may be drawn from the above account appears to support the arguments offered by Braverman (1974) about the degradation of clerical work. Moreover, the growth of part-time employment and the further removing of promotional opportunities provides relatively uncontentious support for dual labour-market theorists. However, neither approach provides an adequate account of the developments described above, primarily because neither provides an adequate account of the impact of product development and innovation. For Braverman, no new work is introduced to alleviate the process of de-skilling because product-markets are implicitly assumed to be mature and stable. For those who stress the role of dual labour-markets as mechanisms for dealing with product-market fluctuation, there is the problem that such institutions initially developed during a period of relatively consistent expansion, both in the banking industry and, more specifically, in the case under consideration. In fact, as in the car industry, change can be understood in terms of the relationship between process innovation, product innovation, and effort bargains. On the one hand, standardized transactions which occur at very high volumes have been subjected to a concern to reduce costs and to maximize volume. This has occurred through increasing levels of counter automation which have reduced several tasks to such simplicity that customers can perform them on auto-tellers, and through work measurement which focuses closely on the counter. The logical end-point for this sort of transaction is thus the removal of all staff input. On the other hand, there are low-volume but high-value transactions which may be supported by the computer system but which require substantial staff time: the best examples here are business loans and business overdrafts,[20] but in general the generation of new account business creates additional work-loads.

This latter set of transactions is difficult to subject to close work measurement: indeed, there may be little purpose in doing so. One reason for this is the nature of the service and of customer involvement in it. As a rule, customer performance remains difficult to control: customers are 'taught' to use ATMs and to complete various forms, but the extent to which they can be compelled to do so is limited by the need to retain custom. So, for example, different branches receive variable work and staff allowances according to the 'slip completion ratio', i.e. the extent to which customers fill in deposit and withdrawal slips correctly. The problem therefore extends to high-volume transactions. On more time-consuming and sensitive transactions, such as loans, customer satisfaction may be

Table 10.6. Work-load and Transaction Times at a Specimen TSB Branch

Item	Average time (mins)	% of work-load[a]
New accounts	16.1	4.2
Account maintenance	2.5	9.8
Account closure	10.7	2.8
New credit	124.1	6.4
Credit services	36.0	3.2
Counter	1.6	33.7
Clearing	0.5	12.0
Mail	4.2	11.8

[a] Main items only: the eight main items account for 80 per cent of work-load.
Source: TSB Branch Establishment Scheme.

related to the amount of advice and consideration which goes along with the business. Moreover, for new products some form of marketing must occur, and once more this is associated with allowances of time—for staff training, business development, and managerial 'public relations'.

The balance between these two sorts of transactions varies between branches: the figures in Table 10.6, which refer to a single branch with an ATM and of about average employment for the ATM sample, are thus merely illustrative but probably not untypical. New accounts and credit business are low volume but high priority; counter, clearing, and mail are the high-volume routine operations. If one takes the transaction time as a rough indicator of skill and responsibility, it is clear that this division of work *may* be paralleled by a division of staff into career–skilled and non-career–unskilled. To the extent that this occurs, it does leave the problem that routine customer contact is with staff whose capacity for and commitment to the sale of other services may be less than that desired.

In fact, account business is performed by cashiers, although most work connected with credit requires more senior staff (grades 3 or 4). Given the division of tasks into these categories, the division of labour is in fact a matter of choice rather than technical necessity. Jobs may be divided between machine-paced short-cycle and long-cycle discretionary tasks, or in such a way as to provide a mix of activity for employees in different grades. The growth in part-time staff strongly implies the choice of the former system of work organization.

The reconciliation of volume with quality concerns thus influenced the search for efficiency through clerical-work measurement. As the most highly automated area, counter work had tended to be the focus of work-study effort. However, the 1983 changes enlarged this to include the improvement of efficiency and an increase in work-rate in back-office areas.[21] The amount of work directly measured was increased from around 30 to 60 per cent of the total. A second change was the shift from a 'task'-based to a 'product'-based work-measurement system. Whereas previous measurement had simply broken down tasks into timed elements, and calculated staffing levels on the basis of timed work-load, the aim of the revamped scheme was to generate a data base for estimation of the costs of providing particular products and services. Marketing would then be able to price products and services more accurately on the basis of knowledge of their staff cost element. The link between changes in product-markets and the organization of work was thus established as part of the work-study scheme itself: product change and development translated directly through to work organization and staffing levels. Since the areas of product growth and development tended to be credit and overseas services, with long average transaction times, and since the process of marketing new products was dealt with by incorporating allowances into the scheme rather than by direct measurements, this could lead in the first instance to longer average times and less measured work, although in the longer term work measurement is likely to be extended to new products and services.

Overall, the picture of the relationship between technology, efficiency, and the effort bargain which emerges is similar to that in Austin Rover. Certain very specific consequences of change such as the reduction in transaction times, the changes to the effort bargain, and the growth in part-time staff emerge, but close inspection reveals that these consequences result from strategic decisions *about* the technology. So the siting of ATMs and decisions about job design and about the deployment of 'spare' capacity freed by technological change are important. In short, the consequences of change are, from the union point of view, in principle negotiable.

COLLECTIVE BARGAINING

Labour relations in TSB have been rather untypical of the banking sector throughout the post-war period. To recap: NUBE was recognized in 1947, and by 1976 had achieved almost 100 per cent membership in grades up to branch manager, supported by an agency shop. BIFU currently has a distinct TSB section, the bank allows two full-time employees to be seconded as union representatives, and the parties are engaged in consultative and negotiating machinery at national and local level. TSB has no staff association, and there is no membership competition as in the clearing banks. Joint consultative arrangements have existed since the Page Report. BIFU has consistently supported changes along Page lines, on the understanding that it would be consulted about them, arguing 'that prime importance must be given to the rights of staff in the TSB movement to be involved completely in the changes which will come about'.[22] Subsequently the process of merger and centralization was dealt with by local consultation, supported by a clear TSB commitment to the avoidance of forced redundancies which was the union's price for co-operation in change.

However, throughout the 1970s no separate bargaining or consultation machinery dealt with the technological changes which introduced the automated counter system. NUBE strategy was to accept change, but to provide job security. In 1978, for example, TSB closed several computer centres to concentrate processing activity around the new second-generation mainframes. Although it has been suggested that several technical options were on offer at this stage, BIFU's concern was primarily in the successful prevention of forced redundancy. The union's approach throughout the period is described in proposals for the 1983 job-security package:

> Unions must be expected to start from a position of 'no redundancies' when looking after the interests of their members. However, all industries will be subjected to structural change from time to time. Not least when new capital equipment is introduced. Normal business decisions will curtail activities of some departments while others are positively encouraged. Redundancy Agreements are acceptable if they are part of a whole job security package.[23]

To some extent, procedures changed to deal with TSB's attempts to specify effort bargains more closely through work measurement in the 1980s. Once more, the guarantee of no forced redundancy was maintained while staffing and work-load calculations were subject to joint review at local level. The 1982 national procedure additionally allowed for disputes over clerical-work measurement and redundancy to be submitted to independent binding arbitration.[24] A second area of change concerned collective agreement coverage of part-time staff: until 1983, national and local representation had excluded part-time clerical staff, and few were in union membership. However, deployment of part-timers was crucial to local discussions about staffing levels and fluctuations. Hence in 1983 the union pushed, successfully, for an extension of the recognition agreement to part-time clericals.

Further developments followed directly from two sets of concerns which achieved prominence in the early 1980s. The first concern was with the consequences of TSB becoming a public limited company with a single centralized administrative hierarchy and a rationalized branch network. The second was with the consequences of technological change. BIFU had been amongst the first unions to develop policies on new technology and to press for new technology agreements. This claim was pressed in TSB in 1981. In response to concern about both sets of changes, TSB, which was reluctant to enter into a technology agreement, revived consultations at national level, the terms of reference of which included the following:

3.2 The introduction of new products, systems and equipment, and the assessment of operational requirements will be the prerogative of the TSB Group, who undertake to consult with the union at the following stages.

3.3 The consultation stages are agreed as:
—concept;
—planning;
—implementation
and prior to each stage, the TSB Group undertakes to provide the Union with information relevant to the project.

3.4 At each consultative stage, matters identified and agreed as relative to terms and conditions of employment will be referred to the appropriate committee for discussion under the procedural agreement. Pending these discussions, the status quo will remain.[25]

Other terms characterized the form as a consultative not a bargaining body, but the issues listed as relevant covered security of employment and of earnings, which were subsequently dealt with in a 1983 Security of Employment Agreement. This agreement preserves the principle that no redundancy be enforced and established the voluntary redundancy mechanisms displayed in Figure 10.6, which require alternative job offers: it, too, extends to part-time employees.

In short, then, TSB has retained collective union influence over major terms and conditions of employment for many years. In par-

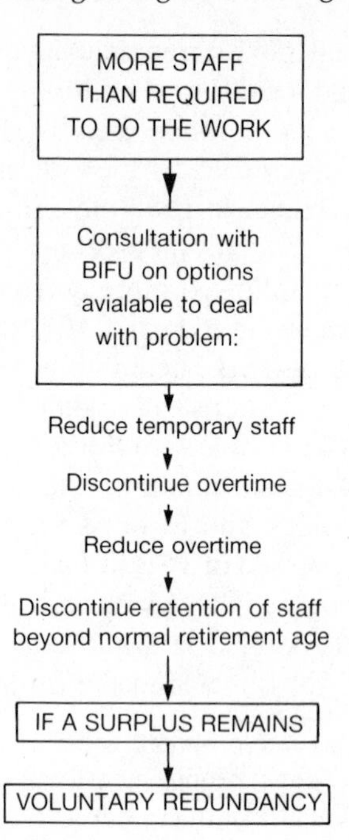

Figure 10.6. TSB Voluntary Redundancy Scheme

ticular, the parties have agreed mechanisms to deal with reorganization and the consequences of new technology. The union did not achieve any form of mutuality, since TSB reserved the right to implement change after discussion rather than after agreement. Nor did it exercise direct control over the terms of the effort bargain or the impact of technology on it. There were no agreements on job content, and the relative lack of concern with it was indicated by its coverage under consultation: 'Where changes in the content of existing jobs emerge which de-skill the job, the current job holders will not suffer any diminution of terms of employment currently enjoyed.'[26] The main form of influence over the level of effort was joint review of work measurement.

However, these developments differed from those elsewhere in banking, and the comparison with the clearing banks is in some ways instructive. BIFU had sought technology agreements in the clearing banks since 1979 without success: in fact, it had failed even to establish consultative arrangements and had not achieved any redundancy guarantees. Over the period 1979–84, the national union policy, heavily influenced by the lack of development in the clearers, had hardened. Beginning with a concern to establish new technology agreements on TUC guide-lines, and to move towards a 28-hour week, the union had increasingly begun to emphasize that side of the policy which followed from employer refusals to negotiate. Eventually, this led to a concern actively to oppose the introduction of new technology and, in 1984, to the endorsement of a motion which recommended resistance to any technological change which led to permanent job loss.[27] The union has experienced compulsory redundancy due to reorganization at Midland Bank, although to date no clerical job loss directly attributable to new technology has occurred. In addition, the individual clearing banks have withdrawn from procedural agreements which required unilateral, binding arbitration.

Differences in approach to consultation and job security mirror differences in other areas. Probably the major 'restrictive practice' in the industry is the restriction of opening hours which has come under increasing pressure as banks seek to improve customer service. BIFU have unsuccessfully opposed Saturday opening elsewhere in the clearers: such practices operate without union agreement. However, TSB has had some success. In April 1985, an agreement was signed between TSB and BIFU securing six-day opening in a large proportion of UK branches: in return, the bank conceded a 32-hour week and an $8\frac{1}{2}$ per cent pay supplement for those involved. In Scotland, a seven-day 28-hour week has been established on an experimental basis. Both are likely to lead to employment expansion.

CONCLUSION

The banking industry has experienced continuous technological change, but the purposes behind it and its consequences have altered over time, as increased competition, particularly in the retail banking area, has tended simultaneously to exert downward pressures on cost and upward pressures on quality and product range. As far as labour relations are concerned, the timing of change has been very important. Quite roughly, one might separate a period of innovation characterized by expansion and poor unionization from a one where growth in staff has reduced or ceased and unions seek influence over the direction and pace of change.

The three phases of change described at the beginning of the section on 'The pattern of change' altered occupational structure, work organization, and employee relations. Computer staff have been introduced as a separate career hierarchy, the branch network has undergone contraction, and the number of non-career cashiering staff, many of them part-time, have grown proportionately. A paternalistic authority relationship gave way to an internal labour-market—at least for males—which is in turn disintegrating. Overall, unions have interpreted recent changes as requiring a low-trust response.

Employers' reactions to this response have differed. Faced with growing concern about employment and careers in general and technological change in particular, clearing banks have chosen not to consult or negotiate. TSB has retained this managerial prerogative together with information-disclosure agreements which mitigate the consequences of change and a willingness to go to binding third-party arbitration. It has thus undertaken many of the measures which, it was argued, were necessary for the avoidance of opportunism.

If the consequences of new technology described here apply more generally, then one could argue that internal labour-markets do face rather severe problems where change is other than incremental. Technical change may lead to a demand for skills which can be accommodated by 'modular' additions to the existing hierarchy—as in the development of 'super-craftsmen' in manufacturing. But where the whole basis of career progression is altered, and career frustrations ensue, employers may find that they are moving from a high-trust to a low-trust effort bargain. The choice appears to be between accepting this as a general trend or confining low-trust relations to a section of the work-froce.

In collective bargaining, one consequence of technical change may be the creation of a new form of trade-unionism. I showed in Chapter 9 how Austin Rover Group sought explicitly to encourage 'constitutional' union activity wherein bargaining was relatively infrequent. However, change can at least have the potential to promote movement in the opposite direction towards local bargaining, as illustrated by TSB, where local bargaining over manning and work standards has developed with technological change: such movement would also be a natural consequence of the disintegration of internal labour-markets.

NOTES

1. Except where indicated to the contrary, the term refers to MLH 861 in the 1968 SIC and 814 in the 1980 SIC.
2. Figures are quoted from the annual *Bankers Magazine* surveys of active ATMs in the retail banking industry.
3. Staff costs in 1980 approximated 70 per cent of total operating costs in the clearing banks compared with about 65 per cent in TSB and about 50 per cent in the building societies. In the 1980s they began to fall as a percentage of overall assets in the clearers but to rise in the other two sets of institutions (Revell 1985, 53–6). The index of average earnings (1976 = 100) stood at 225.9 in June 1982 for all industries and at 247.4 for banking and bill discounting.
4. This discussion is based upon 'Clerical Job Evaluation in the English Clearing Banks', NUBE, 1976.
5. The TSB is an exception here: see the subsection 'Employment composition and the internal labour-market' in this chapter.
6. The increase over the period 1970–8 was over 1000 per cent (TSB, Evidence to the Wilson Committee, Nov. 1978).
7. The initial agreement was for an 'agency' shop under the terms of the 1971 Industrial Relations Act. Similar arrangements were established in the Co-operative and Yorkshire Banks.
8. The two principal sources here are, for the 1973 systems, Wright (1978) and, for subsequent changes, Cowan and Willman (1982). Both are based directly on TSB records.
9. TSB Groups, Business Plans 1983–6.
10. 'Recommended Strategy', TSB Personnel Division Discussion Paper, 1980. Several projections are used based on differing assumptions about business-volume change, but all produce a figure higher than that quoted.
11. The career booklet is entitled 'Come and grow with us', the non-career booklet, 'Come and help us grow'. Both produced by TSB; (n.d.).
12. Work-load is measured in minutes of work: see the subsection 'Technological change and work organization' in this chapter. Source: TSB, Manpower Planning.
13. Variations between 30 and 50 per cent of total work-load were observed: TSB Manpower Plan, 1983.
14. Interview, TSB Management, 18 Sept. 82. The twelve-month training period was originally established by a national agreement (Wright 1978, 89).
15. 'Autoteller Operating Costs: Review', TSB 1982. These two volumes are not necessarily the same, since ATMs may generate additional transactions. A contrasting basis for ATM provision lies in consideration of

the network of ATMs available to the customer on a one-bank or shared basis. On this latter basis, branch location is more important than branch size and the staffing consequences are different.

16. The percentage changes were: full-time, + 1.3 per cent; part-time, + 0.29 per cent.
17. Moreover, no significant relationship existed between percentage change in staff numbers and work-load measures.
18. Correlation between full-time changes, $r = -0.265$; for part-time changes, $r = -0.285$. Both are significant beyond 0.001.
19. Interview, 21 Mar. 1983. ATM transactions had to have a lower average value than counter transactions, and thus had to be more frequent. A *product*—'autotellers'—is built into the work-measurement scheme, and work is generated as the ATM is launched and cards marketed.
20. Whereas, for example, the work-measured transaction time for a cheque-based counter transaction is about 1.5 minutes, for establishing a new business loan it is 683 minutes.
21. Counter work had been rated at 85 per cent BSI and back-office work at 75 per cent BSI. The equalization of these rates at 85 per cent was seen to lead to a reduction of 4.5 per cent in required staff numbers over a two-year period. The plan proposed that 'the saving will be ploughed back and used for business development purposes' (Review of Branch Establishments Scheme, Oct. 1983).
22. NUBE 'Report on Page Recommendations', 1974.
23. BIFU, 'Trustee Savings Bank: Job Security Agreements', n.d.
24. Together with pay, holidays, hours, and health and safety issues.
25. TSB 'Consultative Committee Terms of Reference', n.d.
26. TSB, 'Consultative Committee Terms of Reference', n.d.
27. See 'Microtechnology, A Programme for Action', BIFU, 1981, and Conference *Report*, 1984.

11

Conclusions

SUMMARY

This volume set out to assess the proposition that unions in the UK resisted the introduction of technological change in industry. A review of TUC policy, presented in Chapter 2, illustrated that such organized resistance was not the principal aim of the union movement: however, an analysis of the effects of union policy, both in the UK and the USA, showed that, in the absence of government intervention, trade unions would come to rely upon collective bargaining mechanisms which, at the least, impose adjustment costs.

Trade union policy and practice may thus differ. In Chapter 3 more direct evidence of resistance to change was examined. Although rates of innovation do appear to be lower in the UK than elsewhere, union resistance to innovation did not appear generally to affect such rates or the diffusion of industrial innovations. There was direct evidence of restrictive practices as well as inefficient use of labour due to managerial deficiencies: these problems appeared to be associated with mature industries and large plants. However, the clearest evidence of resistance—strikes in opposition to particular technological changes—has been confined in the last twenty-five years to three industries: docks, national newspapers, and motor vehicles.

Rather than illustrating widespread resistance to change, the evidence revealed that some unions opposed some changes some of the time: in an attempt to go beyond this, the argument proceeded by disaggregation of the main variables. Chapter 4 elaborated the distinction between process and product innovation, and described the relationship between the two within a model of corporate innovative behaviour. Chapter 5 presented a model of different types of collective bargaining within the framework provided by institutional economics. Conflict over change is particularly likely where two conditions exist. On the one hand, cost-cutting process innovations are more likely to generate resistance than product change: expanding sectors where product rather than process change predominates are less likely to experience problems. On the other hand, spot-contracting bargaining structures which transmit product-market volatility through to wages and employment, and which allow continuous union influence over job content, are likely to generate conflict.

All three industries experiencing a high level of conflict over change were characterized by product-market volatility and spot contracting. In all cases, technological change which cut costs was accompanied by organizational change which sought to move away from spot contracting to a more comprehensive contract. The docks were affected by mechanization designed to assist the application of mass-production techniques to a batch operation. In national newspapers, cost-cutting innovation sought to remove anachronistic craft controls from one stage of the production process. The extended comparisons between the UK and USA presented in Chapter 6 reveal that these developments were not peculiarly British. In both countries, conflict emerged from the desire to cut labour costs at all stages of the process: the technology merely offered assistance.

The third industry, motor vehicles, was rather different. Recurrent process and product innovation generated conflict over technological change. However, only part of the UK industry was affected. This part, British owned and reliant on spot contracting, could be contrasted with the US-owned sector of the industry, particularly Ford, as well as with the industry in the USA itself. American Car manufacturers tended towards comprehensive contingent claims contracts which appeared to leave employment and earnings much less susceptible to product-market change.

Given the alleged importance of microprocessor-based process technology, its impact was considered separately. Firms in two industries which have been considerably affected by such process changes, motor vehicles and banking, were selected for closer analysis. In both cases, technical and contractual change coincided as the firms used process changes both to cut costs and to develop new products. As a consequence, certain problems arose. In Austin Rover, authoritarian management and a concern to encourage employee involvement were unhappy bedfellows. At TSB, an internal labour-market, which I have argued is merely a form of contingent claims contract, was in decay. However, the role of collective bargaining as a mechanism for the successful regulation of change differed markedly. At TSB, information disclosure and a job-security agreement helped to retain trade union commitment to both technical and organizational change. At Austin Rover, in very different economic and organizational circumstances, collective bargaining broke down.

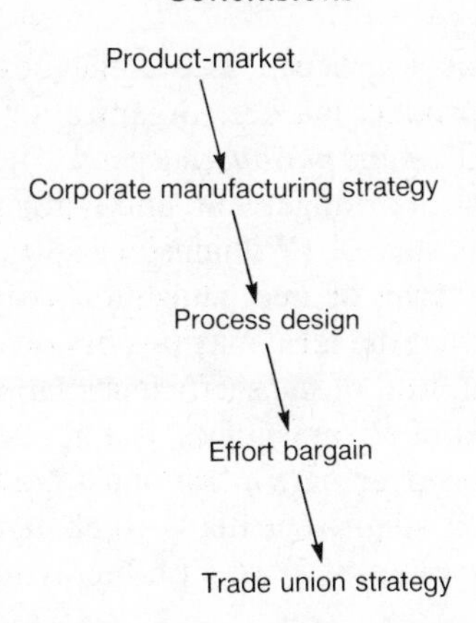

Figure 11.1. Innovation and Trade Union Response

Because technological change is clearly located within corporate strategy, the starting-point of the approach offered here is the corporate response to product-market pressures—either to develop new products or to change the price, quality, or volume of existing ones—rather than the process technology itself (Figure 11.1). Companies are in business to produce goods or services, rather than to run production systems *per se* or to regulate labour relations, and these innovations tend to follow from the requirments of product-market competition, whether on performance or price. Only certain production strategies are feasible within a given product-market, or combination of product-markets, at a given time, and technology will be constrained by the production strategy. Moving further along the chain, different forms of effort bargain are more or less appropriate for different production systems, and trade union reactions to change will depend heavily on the kinds of effort bargains they seek to regulate and the success they have in securing some form of regulation.

At each stage, decision makers have a certain amount of strategic choice: the arrows in Figure 11.1 indicate broad influences rather than determining relationships. As I have shown, different firms may make different choices at each point. Faced with similar product-markets, they may differ in their manufacturing policy, their process design, their mechanisms for labour contracting, and consequently their relationships with trade unions. Nevertheless, decisions made higher up the chain reduce subsequent degrees of freedom, and it is possible—as in the car industry example—to discriminate between firms in terms of the 'goodness of fit' between product-markets and labour relations. Firms which can secure the best match between production process, product-market, organizational structure, and labour relations should, *ceteris paribus*, succeed. The assumption here is that there is one best-fit arrangement, and that a process of natural selection will assure its success (Williamson 1980). However, where efficiency is defined in terms of an organization's capacity to compete and survive, it is clear that the term may possess several meanings. As Silberston notes, the notion of *static* efficiency, applied to the use of resources in a given set of circumstances, is a necessary condition for survival, but for success over time a firm must possess *dynamic* efficiency, in the ability to adapt repeatedly to changing circumstances (1983, 36–7). The example of Ford's problems in the 1920s illustrates the point.

Under different technological and product-market conditions, then, efficiency defined in terms of competitive success may mean very different things. For example, in performance-maximizing industries where competition on quality rather than price prevails, those organizations are 'efficient' which generate successful products, and efficiency thus lies in the success of product design and the quality aspects of production. The idea of efficiency which enjoys the widest currency relates to the engineering definition of mass-production efficiency. Here the level of process technology is seen as the single greatest influence on the productivity potential of a production operation, and efficiency is measured in terms of process yield or the performance of the process against potential. The concept of efficiency employed makes no reference to non-price competition.

These two ideas of efficiency obviously generate very different views concerning the role of direct labour and the operation of effort bargains. The former implies relatively high-trust reliance on employee expertise in the development of new products, whereas the latter implies an attempt to achieve process efficiency almost despite the existence of labour input. Innovation must have very different consequences in the two productive systems. Micro-electronics increases the difficulty, since it implies flexible product strategies in tandem with high process efficiency. Elements of high- and low-trust effort bargains are implied.

Moreover, if one admits a motivational aspect to efficiency, as most institutional economists do, and that different forms of contract have different motivational consequences, the efficiency of particular firms becomes indeterminate in the absence of detailed knowledge about employee reactions to particular contractual forms. The Austin Rover example is particularly relevant here. The search for competitiveness involves using disciplinary measures which are de-motivating to ensure process efficiency while trying to motivate employees to pursue quality by encouraging involvement. The pursuit of involvement may have negative process-efficiency consequences (for example, if the line is halted for some form of communications exercise) which need to be offset against 'x-efficiency' gains.

Since the definition of efficiency may change within a given industry over time, simple generalizations about the superiority of successive institutional arrangements are difficult (Turk 1983, 191–3). Nevertheless, with the above considerations in mind, some broad generalizations are possible. Where product-markets are variable or unpredictable, spot contracting limits the burden carried by the firm, transmitting product-market changes to employees in the form of variable employment and earnings. However, for the reasons discussed in Chapter 5, the choice of this form of contracting actually *creates* a form of trade-unionism in which bargaining strategy develops as an attempt to minimize employment and earnings variation by rules which take advantage—rather than account—of product-market fluctuations. If this form of contracting is retained during a period of expansion, or if product-markets stabilize, firms will need to pay for incremental changes and may well find radical change obstructed.

Where product-markets are mature, production systems with high process efficiency enjoy comparative advantage, and economies of scale are important since competition is primarily on cost. Innovation is absent or incremental, and efficiency in labour utilization is associated with the absence of disruptive haggling over labour supply. A more comprehensive contract which excludes or constrains such haggling is implied: typically this would take the form of a contingent claims contract of limited duration. This contract can offer job security or earnings guarantees for the acceptance of incremental change, and can provide for detailed joint administration of the consequences of managerial decisions while omitting any union influence on such decisions: it is thus a low-trust arrangement which is inappropriate for a period of radical change in process or product.

The problems presented by the impact of micro-electronics in manufacturing are rather more difficult to summarize. The model of process and product innovation suggests that in industries at advanced levels of process efficiency a period of flux or 'de-maturity' will result, with rapid product and process change, and consequently changed labour requirements. This is not merely a statement about the quantity of labour required but also concerns its quality and the type of effort bargain implied. Under such conditions, where competition is on the basis of some combination of quality and price, low-trust contingent claims contracts may break down. As Abernathy *et al*. put it:

> The imperatives of quality and productivity, which lie at the heart of this new industrial competition, are impossible to satisfy without the active, loyal and committed participation of a well trained and constantly improving work-force. In this new environment, what passed for an American labor force policy in previous years is not only out of date: it is poison. (1983, 90.)

The emphasis here shifts to a concern to enhance 'voice' mechanisms, as the Harvard school would put it, or to encourage what Williamson calls 'consummate co-operation'. It would appear to imply a shift to high-trust relations such as that suggested by Sabel, but the example of the car industry, which has passed through these broad contractual phases, suggests a combination of disciplinary measures and involvement exercises.

Other trajectories of change are, of course, possible—particularly outside manufacturing. In banking, for example, an authority relationship gave way over time to an internal labour-market, which currently shows some signs of disintegration. In other services, such as postal work or railways, and indeed in manufacturing industries such as steel, internal labour-markets have existed for a considerable time. The possible combinations of product, process, and effort bargain are numerous. However, in all cases, the form of contract chosen closely defines the form of trade-unionism which emerges, and this in turn defines reactions to technical change.

CONTRACT AND TRADE-UNIONISM

In some ways, the approach I have offered here is similar to that of Clegg in arguing that different forms of union behaviour can be explained in terms of different forms of job regulation or collective bargaining (1976, 4). Both approaches are concerned with the structure and performance of the institutions of job regulation. Moreover, the problems which I have outlined above are similar to those conventionally studied in academic industrial relations. For example, my concerns with the properties of spot contracting, and with its consequences for efficiency in mass-production industries, do not differ markedly from the concerns of those analysing 'disorderly' industrial relations in the 1960s. More specifically, the study of spot contracting in the car industry was couched in terms of the obsolescence of particular institutions, echoing the verdict of Turner *et al*. (1967). However, the reliance on the theorizing of institutional economists does make for several differences. The first is that, whereas the individual relations literature 'is still characterised by fact-finding and description rather than theoretical generalisation' (Winchester 1983, 101), I have sought explictly to generalize both about the factors underlying choice of contractual forms and about their consequences. This has in turn involved movement away from the view that rules of job regulation can be considered in isolation from wider influences to a view which stresses that management choice in the face of product-market constraints strongly conditions developments within the traditional subject area of industrial relations.

Most importantly, the adoption of the organizational failures framework enables two issues to be addressed. One is that of the relative efficiency of different organizational forms in different circumstances. Relatedly, there is the question of how institutions adapt to technical change. Neither is dealt with paticularly well by conventional industrial relations analysis. For example, the view of Turner *et al*. about institutional obsolescence in the car industry relies on the development of sets of employee expectations whose origins remain unclear. I would suggest that an analysis of the properties of different contractual arrangements takes one rather further.

These issues are particularly important for trade union behaviour. The job-regulatory institutions with which unions are involved tend to define the form that union policies and behaviour take. In particular, they affect trade unions' predisposition to oppose change. In spot contracting, for example, trade unions must almost by definition be decentralized: the requisite form of bargaining puts pressure on local negotiators. Given its origin in product-market variation, trade unions are highly likely to become concerned with job control and to seek to bargain over work-loads and employment levels. Incremental technical change is a bargaining opportunity, but contractual change must involve a substantial power shift away from local representatives towards those who will design and administer a comprehensive contract. Day-to-day job control will be forgone. Since the essence of spot contracting is sectional small-group activity, the accommodation of change characteristic of 'encompassing' organizations cannot be achieved and a rather different system for governing the terms of change must be observed. Hence, in the three strike-prone UK industries discussed, government involvement, conflict between shop stewards and national union officials, and conflict between unions were all endemic.

Spot-contracting unionism involves allocative inefficiency. The type of unionism that might generate x-efficiency gains looks quite different. It must provide a communications device (the 'voice' mechanism) between the company and employees, reduce competition and encourage team-working, discourage striking, and establish some form of equity between individual effort bargains. It must receive managerial support and avoid the pursuit of job control since 'If management reacts negatively to collective bargaining or is prevented from organising the work process, unionism can have a negative impact on the performance of the firm' (Freeman and Medoff 1979, 47). Union wage bargaining may have a 'shock' effect in stimulating managers to secure higher output from a given input, but bargaining is to be occasional rather than continuous, and local influence on managerial action is to be avoided. To support the 'voice' aspect, companies might support union membership, and grievance and consultation procedures, seeing the union as an extension of the personnel office (Willman 1980; Terry 1983): a more encompassing form of organization is likely, run from the centre by administration of a complex contract.

These two types involve completely different forms of trade-unionism: they are the 'two faces' described by Freeman and Medoff. While the extremes of aggressive sectional bargaining and what is essentially 'company unionism' are incompatible, most unionized firms probably operate with trade unions which combine aspects of both. If the arguments of this volume are correct, then a firm moving from the first to the second type without offsetting costs will *ceteris paribus* generally enjoy competitive success. This is particularly the case where microprocessor-based innovations are concerned. The successful labour relations strategy appropriate to mass production yields high capacity utilization: that appropriate to new technology yields high capacity utilization *and* consummate co-operation. The technology allows cost reduction, quality improvement, and increased operational flexibility, and these three parameters define successful labour relations strategy.

Technical change may thus lead, through the paths depicted in Figure 11.1 to the creation or destruction of a particular form of trade-unionism. In all the successful cases of change discussed—docks, the car industry, and banks—innovation substantially altered procedural and substantive aspects of the effort bargain in such a way that union organization and policies changed. This raises a number of general issues.

The first has to do with the role of managerial strategy. If companies have the freedom to choose contractual forms, they may effectively influence the form of trade-unionism with which they deal. Secondly, if the different types of trade-unionism have different consequences for productivity and efficiency, then decisions about contractual forms may be important for competitiveness. The third point has to do with the stage of the innovative cycle: the move to mass production is likely to be seen as an unwelcome shift to low-trust relations, but the concern with involvement which may characterize microprocessor-based changes may not be seen as such. Different generations of process change may thus unevenly occasion resistance.

However, technical change may encourage shifts in trade union strategy, particularly in pursuit of membership growth. Process changes which blur distinctions between different types of labour input may, as in printing, encourage union mergers. In banking, where previously distinct financial sectors are entering into competition stimulated by new technology, unions seeking to retain organization of the product-market pursue structural diversification (Willman and Morris 1985). More generally, where 'voice' mechanisms are important to employers, unions may seek to organize on the basis of their commitment to consultation or involvement exercises. All such considerations affect the likelihood of future conflict over change.

TRADE UNION RESISTANCE TO CHANGE: THE FUTURE

The future development of trade union behaviour in the face of technical change will depend upon changes to the political and economic climate, the incidence and impact of innovations, changes to trade union structure, and, perhaps most importantly, collective bargaining developments. I shall consider each in turn.

The disappearance of the political ground from which the TUC launched successive *Economic Reviews* calling for reflation of the economy and a policy of directed investment significantly affects the development of policy towards technical change. *Employment and Technology*, essentially an offshoot of this Keynesian theorizing, withered in the sort of political climate which appears likely to prevail in the medium term. There would seem to be little point in the advocacy of reliance on government to ease the problems of change, and the full weight of adjustment falls upon collective bargaining mechanisms. There also seems little likelihood that the TUC would promote a policy of resistance to innovation: it has never done so, and the manifest futility of such a course of action is likely to prevail over the most persistent demands to do so from affiliates.

The economic climate also points towards the absence of resistance. Since the middle of 1980, output and output per head have risen in the economy as a whole, but unemployment has continued to rise; in manufacturing, output and employment have both fallen sharply over the same period. However, while this is the sort of economic background which promotes mounting concern within the TUC about the effects of new technology, it hardly provides any basis for generalized resistance. Most projections expect no rise in manufacturing employment, but most predict rising productivity, albeit not quite of the rates achieved between 1981 and 1983.

The pattern of innovation implied by this trend is one of cost-reducing process change. One of the complaints raised by writers on the left such as Benson and Lloyd is that the UK has a low international share of the micro-electronics industry: a corollary of this is that the UK economy may, in comparison to those of Japan and the USA, consist of industries which are primarily customers for the products of the performance-maximizing micro-electronics sectors of other nations. If this is the case, then it further follows that innovation shortfalls follow from industrial structure rather than from trade union resistance. Many of the larger equipment suppliers, such as IBM, Hewlett–Packard and Digital, which do operate in the UK, are non-union; moreover, the much publicized drive by the EETPU for no-strike agreements in the electronics sector illustrates the extent to which UK unions are prepared to bend to achieve a foothold in high-technology industries.

The upshot of these considerations is that, *ceteris paribus*, the UK might experience union resistance disproportionate to the amount of change. In particular, if the diffusion of microprocessor-based innovation across the economy is prompted by process-efficiency considerations, and if the high proportion of externally produced innovations deprives UK companies of the ability or disposition to negotiate about it, then a certain amount of resistance might follow.

If the pattern of the past is to continue, then this resistance is most likely to take the form of overt strike action where spot contracting in the presence of bargaining power prevails. As Chapter 6 illustrated, the innovative episode in national newspapers is by no means over: at the time of writing, spot contracting in television is associated with disputes over technological change. Overall, the coverage of such forms of bargaining has almost certainly decreased.

The political climate is particularly important in the public sector. A concern to hold down the level of public-sector expenditure, and in particular a concern with expenditure on wages, may induce cost-cutting process innovations in an environment where competitive pressures may not have been felt. In turn, given the high levels of union organization in the public sector, this may occasion resistance of one form or another. The Civil Service in general, and the Inland Revenue Staff Federation in particular, have unsuccessfully sought new technology agreements. NALGO has achieved agreements in local government, and has been involved in strike activity in opposition to technical change in Sheffield. Events at the Post Office, where strike activity was threatened over new technology and the growth in part-time staff, hint at the decay in the internal labour-market of the sort described at TSB.

However, evidence from the private sector illustrates that the period 1980–4 was one of radical change to working practices and removal of productivity restrictions. There is much myth making in this area, but Incomes Data Services list twenty-seven large organizations which effected improvements in labour efficiency through reform of working practices during the period (IDS 1981, 1984); moreover, most of these companies are outside engineering, an industry in which the reduction in working hours without productivity loss implied reform of working arrangements. In addition, craft flexibility arrangements have been introduced throughout the car industry, as well as in food and chemicals (IDS 1984; Cross 1985). The standard analysis of the development of custom and practice rules, supported by the fact that attempts to improve labour productivity come in regular phases, implies that such attempts do not result in once-and-for-all successes. However, taken with the trends in output and employment, the evidence suggests at the least that many innovating employers may have prevented the extension of restrictive practices to new equipment by removing them altogether.

While the coverage of spot contracting has almost certainly decreased, there appears to be no basis for presuming that contractual forms which would remove the prospect of opportunism have massively expanded in coverage. The relative failure of the 'technology agreement' initiative in the private sector has been discussed above: few companies appear to allow the operation of such agreements. There has been an expansion in consultative-committee coverage including consultation over 'future trends' and 'administrative changes', but one cannot therefore conclude that employees have managed thereby to exert influence on decision making (Daniel and Millward 1984, 129–39).

The final area to consider is that of trade union structure. Given the decline in the number of unions and the rise in their average size, then, on Olsen's analysis, one might expect a marginal reduction in the propensity to resist change which a continuation of the trend over time would enhance. Unions are becoming more 'encompassing associations'.[1] However, in the UK, union size has not noticeably affected the disposition or capacity of certain groups to oppose change: for example, most dockers were in the TGWU, the largest union, during the events described in Chapter 6. However, recent legal changes, which affect the ability of groups within a union to take action in opposition to change, may affect sectional activity. Restriction of the scope of immunity to the 1980 and 1982 Employment Acts, and the balloting regulations of the 1984 Trade Union Act, come to mind. In fact, it is of interest that there has been a technical dimension to several of the disputes which have created legal precedent over recent years. The strike by the POEU over the link with the Mercury system, the *Stockport Messenger* dispute over union membership, and indeed the miners' strike were all to some extent influenced by technical advances in telecommunications, photocomposition, and, in the third case, computerized control of coal-faces which all threatened employment levels in the unions' membership territory. The legal consequences of these various disputes may, however, prevent this recurrence.

Overall, the incidence and impact of technical change in the UK in the near future is likely to generate considerable pressure towards trade union resistance just as the economic, legal, and political climate undercuts the capacity of affected unions to offer it. Product innovation in performance-maximizing areas might occasion a different form of resistance as employees could impose costs without immediate fear of job loss: process innovation designed to cut costs frequently involves the alternatives of acceptance or closure. The paradox facing trade unions is that resistance is most effective where it is least likely, and least effective where the need appears greatest.

For the TUC and its affiliates, current policy options are unclear. The national confederation can rely neither on government action nor on the collective bargaining strength of affiliates to secure peaceful adjustment to change. Affiliates themselves are differentially affected, but the choices faced by unions in contracting manufacturing industries where technological change may be presented as a lifeline, or in service industries threatened by radical process innovation but without the leverage to resist, do not realistically allow for hard bargaining. Nevertheless, the motivation to bargain hard over change has not been removed by institutional development within firms, but rather by allowing unemployment to rise. It is by now something of a truism to say that a change in economic circumstances may see the re-emergence of distributive bargaining over change, but this repetition need not affect the concern on the part of trade unions to attempt in future to right the perceived wrongs of the past.

NOTE

1. According to DE figures, average union size increased from 20 602 in 1970 to 28 541 in 1982.

References

Abernathy, W. J., *The Productivity Dilemma: Roadblock to Innovation in the Auto Industry*, Baltimore: Johns Hopkins, 1978.

Abernathy, W. J., Clark, K., and Kantrow, A., *The New Industrial Renaissance*, New York: Basic Books, 1983.

Abernathy, W. J. and Townsend, P. C., 'Technology, Productivity and Process Change', *Technical Forecasting and Social Change*, 7 (4), 1975, 379–96.

Abernathy, W. J. and Utterback, J. M., 'Patterns of Industrial Innovation', in M. Tushman and W. Moore (eds.), *Readings in the Management of Innovation*, London: Pitman, 1982, 97–109.

Abernathy, W. J. and Wayne, K., 'Limits of the Learning Curve', *Harvard Business Review*, Vol. 52, No. 5, 1974, 109–19.

ACAS, *Industrial Relations in the National Newspaper Industry*, London: HMSO, 1976.

Addison, J., 'The Evolving Debate on Unions and Productivity', *Journal of Industrial Relations*, Vol. 25, No. 3, Sept. 1983, 286–300.

Addison, J. T. and Barnett, A. H., 'The Impact of Unions on Productivity', *British Journal of Industrial Relations*, Vol. 20, No. 2, July 1982, 145–63.

Addison, J. and Siebert, S., *The Market for Labor: An Analytical Treatment*, Santa Monica: Goodyear, 1979.

AFL–CIO, *Adjusting to Automation*, Washington DC: AFL–CIO, 1969.

AFL–CIO, Industrial Union Department, *Comparative Survey of Major Collective Bargaining Agreements*, Washington D.C.: AFL–CIO, 1980.

Altshuler, A., Anderson, M., Jones, D., Roos, D., and Womack, J., *The Future of the Automobile*, London: George Allen and Unwin, 1984.

Arnold, E. and Senker, P., 'Designing the Future: The Implications of CAD for Employment and Skills in the British Engineering Industry', EITB Occasional Paper 9, 1982.

ASTMS, *Technological Change and Collective Bargaining*, London: Association of Scientific, Technical, and Managerial Staffs, 1980.

Ayres, R. O. and Miller, S., 'Robotics, CAM and Industrial Productivity', *National Productivity Review*, Vol. 1, 1981, 42–60.

Ayton, J., 'Plant Size and Efficiency in the Steel Industry: An International Comparison', *National Institute Economic Review*, No. 100, May 1982, 65–76.

Bacon, R. W. and Eltis, W. A., *The Age of US and the UK Machinery*, London: National Economic Development Office, 1974.

Bain, G. S. and Price, R., *Profiles of Union Growth*, Oxford: Blackwell, 1980.

Baldamus, W., *Efficiency and Effort*, London: Tavistock, 1961.

Ball, J. M. and Skeoch, N. K., 'Interplant Comparisons of Productivity and Earnings', DE Working Paper, No. 3, 1981.

Bamber, G. and Willman, P., 'Technological Change and Industrial Relations in Britain', *Bulletin of Comparative Labor Relations*, No. 12, 1983, 101–21.

Batstone, E., *Working Order*, Oxford: Blackwell, 1984.

Batstone, E., Boraston, I, and Frenkel, S., *Shop Stewards in Action*, Oxford: Blackwell, 1977.

Batstone, E., Ferner, A., and Terry, M., *Consent and Efficiency: Labour Relations and Management Strategy in the State Enterprise*, Oxford: Blackwell, 1984.

Behrend, H., 'The Effort Bargain', *Industrial and Labor Relations Review*, Vol. 10, 1957.

Bell, R. M., 'The Behaviour of Labour, Technical Change and the Weakness of British Manufacturing', SPRU mimeo, 1983.

Benson, I. and Lloyd, J., *New Technology and Industrial Change*, London: Kogan Page, 1983.

Bessant, J. R., *Microprocessors in Manufacturing Processes*, London: Policy Studies Institute, 1982a.

Bessant, J. R., 'Influential Factors in Manufacturing Innovation', *Research Policy*, Vol. 11, 1982b, 117–32.

Beynon, H., *Working for Ford*, Harmondsworth: Penguin, 1973.

BIFU, *New Technology in Banking, Insurance and Finance*, London: Banking, Insurance, and Finance Union, 1983.

Black, A. P., 'Long Run Theories of Economic Growth, with Reference to the American and British Automobile Industries', Ph.D. thesis, London University, 1980.

Blackburn, R., *Union Character and Social Class*, London: Batsford, 1967.

Blitz, H. J., *Labor—Management Contracts and Technological Change: Case Studies and Contract Clauses*, New York: Praeger, 1969.

Bloomfield, G., *The World Automotive Industry*, London: David and Charles, 1978.

BLS, 'Characteristics of Major Collective Bargaining Agreements', Bulletin 2005, Washington DC: Department of Labor, 1980.

BNA, *Basic Patterns in Union Contracts*, Washington DC: Bureau of National Affairs, 1979.

Bowers, J., Deaton, D., and Turk, J., *Labour Hoarding in British Industry*, Oxford: Blackwell, 1982.

Brady, T. and Liff, S., *Monitoring New Technology and Employment*, Manpower Services Commission, London: HMSO June 1983.

Braverman, H., *Labor and Monopoly Capital*, New York: Monthly Review Press, 1974.

de Bresson, C. and Townsend, J., 'Notes on the Inter-industrial Flow of Technology in Post-war Britain', *Research Policy*, Vol. 7, 1978, 48–60.

de Bresson, C. and Townsend, J., 'Multivariate Models for Innovation—Looking at the Abernathy–Utterback Model with Other Data', *Omega*, Vol. 9, No. 4, 1981, 429–36.

Bright, J. R., *Automation and Management*, Harvard University: Graduate School of Business Administration, 1958.

Brown, W. A., *Piecework Bargaining*, London: Heinemann, 1973.

Brown, W. A. (ed.), *The Changing Contours of British Industrial Relations*, Oxford: Blackwell, 1981.

Buchanan, D. and Boddy, D., *Organisations in the Computer Age*, Farnborough: Gower, 1983.

BTDB, 'Containerisation: The Key to Low-cost Transport', report by McKinsey and Co. for the British Transport Docks Board, London, 1967.

Burns, T. and Stalker, G. M., *The Management of Innovation*, London: Tavistock, 1961.

Caves, R. (ed), *Britain's Economic Prospects*, Washington DC: Brooking Institution 1968.

Caves, R., 'Productivity Differences Among Industries', in R. Caves and L. Krause (eds.), *Britain's Economic Performance*, Washington DC: Brooking Institution, 1980, 135–92.

CBI, *Jobs: Facing the Future*, London: Confederation of British Industry, 1980.

Channon, D., *The Strategy and Structure of British Enterprise*, London: Macmillan, 1973.

Chell, R., 'BL Cars Ltd—The Frontier of Control', MA thesis, University of Warwick, 1980.

Child, J., 'Organisation Structure, Environment and Performance: The Role of Strategic Choice', *Sociology*, Vol. 6, 1972, 1–22.

Child, J., *Organization*, London: Harper and Row, 1984.

Clark, J., 'A Model of Embodied Technical Change and Employment', SPRU mimeo, 1979.

Clark, J., *et al*. *Trade Unions, National Politics and Economic Management: A Comparative Study of the TUC and DGB*, London: Anglo-German Foundation, 1980.

Clegg, H., *Trade Unionism under Collective Bargaining*, Oxford: Blackwell, 1976.

Clegg, H. A., *The Changing System of Industrial Relations in Great Britain*, Oxford: Blackwell, 1979.

Cockroft, J., 'Microtechnology in Banking', Economist Intelligence Unit, Special Report, No. 169, London, 1984.

Corina, J., 'Trade Unions and Technological Change', in S. Macdonald, D. McL. Lamberton, and T. D. Manderville (eds.), *The Trouble with Technology*, London: Frances Pinter, 1983.

Cowan, N., 'Effects on Staff and Organisation', in *The Banks and Technology in the 1980s*, London: Institute of Bankers, 1982.

Cowan, R. and Willman, P., 'Technological Change and Personnel Policies within the Trustee Savings Banks', report to TSB Central Board, mimeo, Imperial College, 1982.

CPRS, *The Future of the British Car Industry*, London: HMSO, 1975.

Crompton, R. and Jones, G., *White Collar Proletariat: De-skilling and Gender in Clerical Work*, London: Macmillan, 1984.

Cross, M., *Towards the Flexible Craftsman*, London: Technical Change Centre, 1985.

Crouch, C., *Class Conflict and the Industrial Relations Crisis*, London: Heinemann, 1977.

Crouch, C., *The Politics of Industrial Relations*, London: Fontana/Collins, 1979.

Crouch, C., *Trade Unions: The Logic of Collective Action*, London: Fontana, 1982.

Currie, R., *Work Study*, London: Pitman, 1977.

Daniel, W. and Millward, N., *Workplace Industrial Relations in Britain*, London: Heinemann, 1984.

Davies, S., *The Diffusion of Process Innovations*, Cambridge University Press, 1979.

Denison, E. F., *Why Growth Rates Differ*, Washington DC: Brooking Institution, 1967.

Doeringer, P. and Piore, M., *Internal Labor Markets and Manpower Analysis*, Lexington: D. C. Heath, 1971.

Donaldson, L., 'Organizational Design and the Life Cycles of Products', *Journal of Management Studies*, Vol. 21, No. 1, 1985, 25–37.

Durcan, J., McCarthy, W. E. J., and Redman, G., *Strikes in Post-War Britain*, London: George Allen and Unwin, 1983.

Edwards, P. K., 'The Management of Productivity', Warwick IRRU mimeo, 1984.

Edwards, R., *Contested Terrain*, London: Heinemann, 1979.

Egan, A., 'Women in Banking: A Study in Inequality', *Industrial Relations Journal*, Vol. 13, No. 3, 1982.

Fadem, J. A., 'Technological Change, Worker Attitudes and the Employment Relationship in the British Docks Industry', M.Litt. thesis, Oxford University, 1976.

Flanders, A., *The Fawley Productivity Agreements: A Case Study of Management and Collective Bargaining*, London: Faber and Faber, 1964.

Fleck, J., 'The Diffusion of Robots in British Manufacturing Industry', Final Report to the SERC, Aston University, 1982.

Ford, P. W., 'Changing the Wage Payments System in a Car Factory', MA thesis, University of Warwick, 1972.

Fox, A., *Beyond Contract: Work, Power and Trust Relations*, London: Faber and Faber, 1974.

Francis, A., Turk, J. and Willman, P., *Power, Efficiency and Institutions: A Critical Appraisal of the Markets and Hierarchies Paradigm*, London: Heinemann, 1983.

Freeman, C., *The Economics of Industrial Innovation*, Harmondsworth: Penguin, 1976.

Freeman, C., 'Technological Innovation and British Trade Performance', in F. Blackaby (ed.), *Deindustrialisation*, London: Heinemann/NIESR, 1979, 56–73.

Freeman, C., 'Some Economic Implications of Microelectronics', in D. Cohen (ed.), *Agenda for Britain: Micro Policy Choices*, Oxford: Philip Allan, 1982.

Freeman, C., Clark, J. and Soete, L., *Unemployment and Technical Innovation*, London: Frances Pinter, 1982.

Freeman, R. B. and Medoff, J. L., 'The Two Faces of Unionism', *The Public Interest*, Fall 1979, 69–95.

GAO, 'Advances in Automation Prompt Concern over Increased U.S. Unemployment', General Accounting Office AFMD 82–44, Washington DC: GAO, 1982.

Gaskin, B. and Gaskin, M., *Employment in Insurance, Banking and Finance in Scotland*, Scottish Economic Planning Department, 1980.

Giersch, H. and Wolter, F., 'Towards an Explanation of the Productivity Slowdown: An Acceleration–Deceleration Hypothesis', *Economic Journal*, Vol. 93, 1983, 35–55.

Goldberg, J. P., 'Longshoremen and the Modernisation of Cargo Handling in the US' *International Labor Review*, Vol. 107, No. 3, Mar. 1973.

Gomulka, S., 'Britain's Slow Industrial Growth: Increasing Inefficiency versus Low Rate of Technological Change', in W. Beckerman (ed.), *Slow Growth in Britain: Causes and Consequences*, Oxford: Clarendon Press, 1979, 166–94.

Gordon, D. M., 'Capitalist Efficiency and Socialist Efficiency', *Monthly Review*, Vol. 23, No. 3, 1976, 19–39.

Gospel, H. and Willman, P., 'Disclosure of Information: The CAC Approach', *Industrial Law Journal*, Vol. X, 1981, 10–22.

Griffin, A., 'Technological Change and Craft Control in the Newspaper Industry: An International Comparison', *Cambridge Journal of Economics*, 1984, Vol. 8, 41–61.

Hartley, J., *Management of Vehicle Production*, London: Butterworth, 1981.

Hartman, P., *Collective Bargaining and Productivity: The Longshore Mechanisation Agreement*, Berkeley: University of California Press, 1969.

Hax, A. C., 'A Methodology for the Development of a Human Resource Strategy', Working Paper, No. 1638–85, Sloan School of Management, Massachusetts Institute of Technology, 1985.

Hill, C. T. and Utterback, J. M., *Technological Innovation for an Expanding Economy*, Oxford: Pergamon, 1979.

Hirschmann, A., *Exit, Voice and Loyalty*, Cambridge: Harvard University Press, 1970.

Hotz, B., 'Productivity Differences and Industrial Relations Structures', *Labour and Society*, Vol. 7, No. 4, 1982, 334–54.

House of Commons Expenditure Committee, Fourteenth Report, *The Motor Vehicle Industry*, London: HMSO, 1975.

Hutton, S. T. and Lawrence, T. A., *German Engineers: The Anatomy of a Profession*, Oxford University Press, 1981.

Hyman, R., *Disputes Procedures in Action: A Study of the Engineering Industry Disputes Procedure in Coventry*, London: Heinemann, 1972.

Hynds, E., *American Newspapers in the 1970s*, New York: Hastings House, 1975.

IDS, 'Productivity Improvements', IDS Study No. 245, Incomes Data Services, July 1981.

IDS, 'Craft Flexibility', IDS Study No. 322, Incomes Data Services, dept. 1984.

IOB, 'The Banks and Technology in the 1980s', Cambridge Seminar, Institute of Bankers, 1982.

Jackson, M., *Labour Relations on the Docks*, Lexington: Saxon House, 1973.

Jacobs, E., *et al.*, *The Approach to Industrial Change in Britain and West Germany*, London: Anglo-German Foundation, 1978.

Jansenn, J. O. and Schneersen, D., *Port Economics*, London: MIT Press, 1982.

Jensen, V. H., *Strife on the Waterfront: The Port of New York Since 1945*, Ithaca: Cornell University Press, 1974.

Johnson, K. and Garnett, H., *The Economics of Containerisation*, London: George Allen and Unwin, 1971.

Jones, D. T., 'Maturity and Crisis in the European Car Industry', Sussex European Papers, No. 8, Sussex European Research Centre, 1981.

Jones, D. T., 'Technology and the UK Automobile Industry', *Lloyds Bank Review*, No. 148, 1983.

Jones, D. T., 'The Future of the Motor Industry in the UK Economy', unpublished paper, SPRU, 1984.

Jones, D. T. and Prais, S. J., 'Plant Size and Productivity in the Motor Industry: Some International Comparisons', *Oxford Bulletin of Economics and Statistics*, Vol. 40, May 1978.

Jorgenson, D. W. and Griliches, Z., 'The Explanation of Productivity Changes' *Review of Economic Studies*, Vol. 34, July 1967, 249–83.

Jurgens, U., Dohse, K., and Malsch, T. 'New Production Concepts in West German Car Plants', in S. Tolliday and J. Zeitlin, (eds.), *Between Fordism and Flexibility: The International Automobile Industry and its Workers*, Oxford: Polity Press, 1986.

Kamien, M. I. and Schwartz, N. C., *Market Structure and Innovation*, Cambridge University Press, 1982.

Katz, H., 'The US Automobile Collective Bargaining System in Transition', *British Journal of Industrial Relations*, Vol. XXII, No. 2, 1984, 205–18.

Katz, H., *Shifting Gears: Changing Labor Relations in the US Automobile Industry*, Cambridge: MIT Press, 1985.

Katz, H. and Karl, R., 'Personnel Planning in the US Auto Industry', in W. Streek and A. Hoff (eds.), *Workforce Restructuring, Manpower Management and Industrial Relations in the World Auto Industry*, Berlin: International Institute of Management, 1983.

Kelber, H. and Schlesinger, C., *Union Printers and Controlled Automation*, New York: Free Press, 1967.

Killingsworth, C. C., 'The Modernisation of West Coast Longshore Work Rules', *Industrial and Labor Relations Review*, Vol. 15, 1962, 295–306.

Kirchner, E. and Hewlett, N. 'The Social Implications of New Technology in the Banking Sector', mimeo, University of Essex, 1983.

Knott, R. D. and Williams, D. J., 'Containerisation, Unitization and Packaging', in F. Wentworth (ed.), *Handbook of Physical Distribution Management*, Farnborough: Gower, 1976, 281–301.

Kochan, T., McKersie, R., and Capelli, P., 'Strategic Choice and Industrial Relations Theory', *Industrial Relations*, Vol. 23, No. 1, 1984, 16–39.

Kravis, I., 'A Survey of International Comparisons of Productivity', *Economic Journal*, Vol 86, 1976, p 1–39.

Lawrence, P. R. and Lorsch, J. W., *Developing Organisations: Diagnosis and Action*, New York: Addison-Wesley, 1969.

Lee, T., 'Computers in Banking: The London Clearing Banks', *The Banker*, Apr. 1973.

Leibenstein, H., *Beyond Economic Man: A New Foundation for Microeconomics*, Cambridge: Harvard University Press, 1976.

Levinson, H. M., Rehmus, C., Goldberg, J. P. and Kahn, M., *Collective Bargaining and Technological Change in American Transportation*, Evanston: Northwestern University Transport Center, 1971.

Lewchuck, W., 'Fordism and British Motor Car Employees', in H. Gospel and C. R. Littler, (eds.), *Managerial Strategies and Industrial Relations: An Historical and Comparative Study*, London: Heinemann, 1983.

Lewchuk, W., 'The Origins of Fordism and Alternative Strategies: Britain and the United States, 1880–1930', paper to Coventry Car Workers' Conference, mimeo, July 1984.

Lindbeck, A., 'The Recent Slowdown of Productivity Growth', *Economic Journal*, Vol. 93, 1983, 13–34.

Littler, C., 'A History of "New" Technology', in G. Winch (ed). *Information Technology in Manufacturing Processes*, London: Rosendale, 1983.

Littler, C. and Salaman, G., 'Bravermania and Beyond: Recent Theories of the Labour Process', *Sociology*, Vol. 16, No. 2, 1982, 251–70.

Lorsch, J., *Product Innovation and Organisation*, London: Macmillan, 1965.

Lyddon, D., 'Workplace Organization in the British Car Industry: A Critique of Jonathan Zeitlin', *History Workshop Journal*, No. 15, 1983.

McCarthy, W. E. J., 'The Reasons for Striking', *Bulletin of the Oxford University Institute of Statistics*, Feb. 1959.

McCormick, B., 'Methods of Wage Payment, Wages Structures and the Influence of Factor and Product Markets', *British Journal of Industrial Relations*, Vol. XV, No.2, 1977, 246–65.

Macdonald, R., *Collective Bargaining in the Automobile Industry*, New Haven: Yale University Press, 1963.

McKersie, R. B. and Hunter, L., *Pay, Productivity and Collective Bargaining*, London: Macmillan 1973.

McKersie, R. B. and Klein, J., 'Productivity: The Industrial Relations Connection', Working Paper No. 1376-82, Sloan School of Management, Massachusetts Institute of Technology, 1982.

McLaughlin, D. B. and Miller, C. E., *The Impact of Labor Unions on the Rate and Direction of Technical Change*, Ann Arbor: Institute of Labor and Industrial Relations, University of Michigan, for National Science Foundation, 1979.

Maddison, A., 'The Long Run Dynamics of Productivity Growth', in W. Beckerman, (ed.), *Slow Growth in Britain: Causes and Consequences*, Oxford: Clarendon Press, 1979, 194–213.

Mansfield, R., *Industrial Research & Technological Innovation*, London: Longmans, 1968.

Manwaring, A., 'The Trade Union Response to New Technology', *Industrial Relations Journal*, Vol. 12, No. 4, 1981, 7–26.

Marglin, S., 'What Do Bosses Do?', in A. Gorz (ed.), *The Division of Labor*, Brighton: Harvester Press, 1976.

Marsden, D., Morris, T., Willman, P., and Wood, S., *The Car Industry: Labour Relations and Industrial Adjustment*, London: Tavistock, 1985.

Martin, R., *TUC: The Growth of a Pressure Group, 1968—1976*, Oxford: Clarendon Press, 1980.

Martin, R., *New Technology and Industrial Relations in Fleet St.*, Oxford University Press, 1981.

Maxcy, G. and Silberston, Z. A., *The Motor Industry*, London: George Allen and Unwin, 1959.

Meidan, A., *Bank Marketing Management*, London: Macmillan, 1984.

Mellish, M., *The Docks After Devlin*, London: Heinemann Educational Books, 1973.
Melman, S., *Decision Making and Productivity*, Oxford: Blackwell, 1958.
Morris, T., 'The Establishment and Operation of National Negotiating Machinery in the London Clearing Banks', Ph.D. thesis, University of London, 1984.
Mowery, F. and Rosenberg, N., 'The Influence of Market Demand on Innovation: A Critical Review of Some Recent Empirical Studies', *Research Policy*, Vol. 8, 1979, 102–53.
Murphy, R., *Technological Change Clauses in Collective Bargaining Agreements*, Department for Professional Employees, AFL–CIO, Publication No. 81–2, Washington DC: AFL–CIO, 1981.
Nabseth, L. and Ray, G. F., *The Diffusion of New Industrial Processes: An International Study*, Cambridge University Press, 1974.
NCTAEP, *Technology and the American Economy*, Report of the National Commission on Technology, Automation, and Economic Progress, Vol. 1, Washington DC: US Printing Office, Feb. 1966.
NEDO, *International Price Competitiveness, Non-Price Factors and Economic Performance*, London: National Economic Development Office, 1977.
NEDO, *Policy for the UK Information Technology Industry*, London: National Economic Development Office, 1983.
Negrelli, S., 'The Robogate System at FIAT', paper to International Automobile Conference, Berlin, 1984.
Northcott, J., Rogers, P. and Zeitlinger, A., *Microelectronics in Industry: Survey Statistics*, London: Policy Studies Institute, 1982.
Northcott, J. and Rogers, P., *Microelectronics in British Industry: Patterns of Change*, Policy Studies Institute, 1984.
NPC, *Containerisation on the North Atlantic*, London: National Ports Council, 1967.
NPC, *Containerisation, a Survey of Current Practice*, London, National Ports Council, 1978.
NUJ, *Journalists and New Technology*, London: National Union of Journalists, 1978.
OECD, *The Future of the World Automobile Industry*, Paris: Organization for Economic Co-operation and Development, 1983.
Offe, C., *Industry and Inequality*, London: Arnold, 1976.
Olsen, M., *The Rise and Decline of Nations*, Harvard University Press, 1982.
OU, *Press, Papers and Print: A Case Study*, Open University Press, 1976.
Owen, N., *Economies of Scale, Competitiveness and Trade Patterns within the European Community*, Oxford University Press, 1983.
PACTEL, *Automation in Banking*, London: PA Consultants, 1980.
Paige, D. and Bombach, G., *A Comparison of National Output and Produc-*

tivity of the United Kingdom and United States, Paris: Organization for Economic Co-operation and Development, 1959.

Palmer, L. S., 'Technical Change and Employment in Banking', M.Sc. thesis, University of Sussex, 1980.

Panic, M., 'Why the UK's Propensity to Import is High', *Lloyds Bank Review*, No. 115, 1975, 1–13.

Panic, M. and Rajan, A. H., *Product Changes in Industrial Countries' Trade: 1955—68*, London: National Economic Development Office, 1971.

Pavitt, K., (ed.), *Technical Innovation and British Economic Performance*, London: Macmillan, 1980.

Pavitt, K., 'Characteristics of Innovative Activity in British Industry', *Omega*, vol. 11, No. 2, 1983, 113–30.

Pavitt, K., 'Sectoral Patterns of Technical Change: Towards a Taxonomy and a Theory', *Research Policy*, Vol. 13, 1984, 343–73.

Pavitt, K. and Soete, L., 'Innovative Activities and Export Shares: Some Comparisons between Industries and Countries' in K. Pavitt (ed), *Technical Innovation and British Economic Performance*, London: Macmillan, 1980.

PEP, *Motor Vehicles*, London: Political and Economic Planning, 1950.

Piore, M., 'The Impact of the Labour Market on the Design and Selection of Productive Techniques within the Manufacturing Plant', *Quarterly Journal of Economics*, Vol. 821, No. 4, 1968, 602–20.

Piore, M., 'The Dual Labour Market: Theory and Implications', in D. M. Gordon (ed.), *Problems of Political Economy*, Lexington, 1971.

Piore, M., *Birds of Passage*, Cambridge University Press, 1979.

Pollard, S., *The Wasting of the British Economy*, London: Croom Helm, 1982.

Prais, S. J., 'Comment', in R. Caves and L. Krause (eds.), *Britain's Economic Performance*, Washington DC: Brooking Institution, 1980, 193–9.

Prais, S. J., *Productivity and Industrial Structure*, Cambridge University Press, 1981.

Pratten, C. F., *Economies of Scale in Manufacturing Industry*, Cambridge University Press, 1971.

Pratten, C. F., 'Labour Productivity Differentials within International Companies', DAE Occasional Paper 50, Cambridge University Press, 1976.

Pratten, C. F. and Atkinson, A. G., 'The Use of Manpower in British Manufacturing Industry', *Employment Gazette*, July 1976, 571–7. By permission of the Controller, HMSO and the Editor of *Employment Gazette*.

Price Commission, 'Banks: Charges for Money Transmission Services', House of Commons Paper No. 337, London: HMSO 1978.

Price, R. and Bain, G. S., *Profiles of Union Growth*, Oxford: Blackwell, 1980.

Price, R., and Bain, G., 'Union Growth in Britain: Retrospect and Prospect', *British Journal of Industrial Relations*, Vol.'XXI, 1983, 46–68.

Purcell, J., *Good Industrial Relations: Theory and Practice*, London: Macmillan, 1981.

Rajan, A., *New Technology and Employment in the Financial Services Sector*, Aldershot: Gower, 1984.

Ray, G. F., 'The Diffusion of New Technology: A Study of Ten Processes in Nine Industries', *National Institute Economic Review*, No. 48, 1969, 40–83.

Ray, G. F., 'Industrial Labour Costs, 1971–83', *National Institute Economic Review*, No. 110, 1984a, 62–8.

Ray, G. F., 'The Diffusion of Mature Industrial Processes', NIESR Occasional Paper No. XXXVI, Cambridge University Press, 1984b.

Reddaway, W. B., 'The Economics of Newspapers', *Economic Journal*, Vol. 73, 1963, 201–18.

Revell, J., *Costs and Margins in Banking: An International Survey*, Paris: Organization for Economic Co-operation and Development, 1980.

Revell, J., *Banking and Electronic Funds Transfer*, Paris: Organization for Economic Co-operation and Development, 1983.

Revell, J., *Costs and Margins in Banking: Statistical Supplement 1978—82*, Paris: Organization for Economic Co-operation and Development, 1985.

Rhys, D. G., *The Motor Industry: An Economic Survey*, London: Butterworth, 1972.

Rhys, D. G., 'Employment, Efficiency and Labour Relations in the British Motor Industry', *Industrial Relations Journal*, Vol. 5, No. 2, 1974, 4–24.

Rhys, D. G., 'European Mass Producing Car Makers and Minimum Efficient Scale: A Note', *Journal of Industrial Economics*, Vol. 24, June 1977, 313–19.

Robertson, J. A., *et al.*, *Structure and Employment Prospects of the Service Industries*, Department of Employment, 1982.

Robins, K. and Webster, F., 'New Technology: A Survey of Trade Union Response in Britain', *Industrial Relations Journal*, Vol. 13, No. 1, 1982, 7–27.

Rogers, T. and Friedman, N. S., *Printers Face Automation: The Impact of Technology on Work and Retirement Among Skilled Craftsmen*, Lexington: Lexington Books/D.C. Heath, 1980.

Rostas, L., 'Comparative Productivity in British and American Industry', NIESR, Occasional Paper No. 13, Cambridge University Press, 1948.

Rothwell, R., 'The Relationship between Technical Change and Economic Performance in Mechanical Engineering', in B. J. Baker (ed.), *Industrial Innovation*, London: Macmillan, 1979.

Rothwell, R. and Zegveld, W., *Reindustrialisation and Technology*, London: Longman, 1985

Routledge, P., 'The Dispute at Times Newspapers Ltd.: A. View from Inside', *Industrial Relations Journal*, Vol. 10, No. 4, 1979, 5–9.

Roy, A. D., 'Labour Productivity in 1980—An International Comparison', *National Institute Economic Review*, No. 101, Aug. 1982, 26–38.

Royal Commission on the Press, 'Interim Report', Cmnd 6433, London: HMSO, 1976.

Royal Commission on the Press, 'Final Report', Cmnd 6810, London: HMSO, 1977.

Rubery J. *et al.*, 'Industrial Relations Issues in the 1980's: An Economic Analysis', in M. Poole *et al* (eds.), *Industrial Relations in the Future*, London: Routledge and Kegan Paul, 1984.

Ryder Committee, *British Leyland, The Next Decade*, London: HMSO, 1975.

Sabel, C., *Work and Politics*, Cambridge University Press, 1982.

Salter, W. E. G., *Productivity and Technical Change*, Cambridge University Press, 1960.

Schultz, G. P. and Weber, A. R., *Strategies For Displaced Workers*, Westport: Greenwood Press, 1966.

Shaw, E. R. and Coulbeck, N. S., *UK Banking Prospects in the Competitive 1980s*, London: Staniland Hall, 1983.

Silberston, Z. A., 'The motor Industry, 1955–1964', *Oxford Bulletin of Economics and Statistics*, Vol. 27 (4), 1965, 253–86.

Silberston, A., 'Efficiency and the Individual Firm', in D. Shepherd, J. Turk, and A. Silberston (eds.) *Microeconomic Efficiency and Macroeconomic Performance*, Oxford: Philip Allan, 1983, 30–46.

Simmonds, W. H. C., 'Towards an Analytical Industry Classification', *Technological Forecasting and Social Change*, Vol. 4, 1973, 375–85.

Sisson, K. *Industrial Relations in Fleet Street*, Oxford: Blackwell, 1975.

Skinner, W., 'The Focused Factory', *Harvard Business Review*, Vol. 52, May/June 1974, 110–16.

Slichter, S. H., *et al.*, *The Impact of Collective Bargaining on Management*, New York: Brooking Institute, 1960.

Smith, A. *Goodbye Gutenberg: The Newspaper Revolution of the 1980s*, Oxford University Press, 1980.

Smith, A. D., Hitchens, D. M. W. N., and Davies, S. W. 'International Industrial Productivity: A Comparison of Britain, America and Germany', *National Institute Economic Review*, No. 101, Aug. 1982, 13–26; also 'International Industrial Productivity: A Comparison of Britain, America and Germany', National Institute of Economic and Social Research Occasional Paper No. XXXIV, Cambridge University Press, 1982.

Solow, R. M., 'Technical Change and the Aggregate Production Function', *Review of Economics and Statistics*, Vol. 39, 1957, 312–20.

Somers, G., *et al.*, *Adjusting to Technological Change*, New York: Harper and Row, 1963.

Sorge, A., Hartman, G., Warner, M., and Nicholas, I., *Microelectronics and Manpower in Manufacturing*, Farnborough: Gower, 1983.

Stoneman, P., *The Economics of Technological Change*, Oxford University Press, 1983.

Streeck, W., *Industrial Relations in W. Germany*, London: Heinemann/PSI, 1984.

Streeck, W. and Hoff, A. (eds.), *Workforce Restructuring, Manpower Management and Industrial Relations in the World Auto Industry*, Berlin: International Institute of Management, 1983.

Sykes, A., 'The Cohesion of a Trade Union Workshop Organisation', *Sociology*, Vol. 1, 1967.

Terry, M., 'The Inevitable Growth of Informality', *British Journal of Industrial Relations*, Vol. 15, 1977, 75–90.

Terry, M., 'Shop Stewards Through Expansion and Recession', *Industrial Relations Journal*, Vol. 14, No. 3, 1983, 49–58.

Thomas, D., 'Strategy is Different in Service Business', *Harvard Business Review*, Vol. 56, 1978.

Thomson, A. W. J., 'Unions and the Corporate State in Britain', *Industrial and Labor Relations Review*, Vol. 33, No. 1, Oct. 1979, 36–52.

Thomson, A. W. J., 'A View from Abroad', in J. Stieber, R. B. McKersie, and D. Q. Mills (eds.), *US Industrial Relations 1950–80: A Critical Assessment*, Madison: Industrial Relations Research Association, University of Wisconsin, 1981.

Tolliday, S., 'High Tide and After: Coventry Engineering Workers and Shopfloor Bargaining', 1945–80, in S. Tolliday and J. Zeitlin (eds.), *Between Fordism and Flexibility: The International Automobile Industry and its Workers*, Oxford: Polity Press, 1986.

Tolliday, S. and Zeitlin, J., 'Shopfloor Bargaining, Contract Unionism and Job Control: An Anglo American Comparison', in S. Tolliday and J. Zeitlin (eds.), *Between Fordism and Flexibility: The International Automobile Industry and its Workers*, Oxford: Polity Press, 1986.

Townsend, J. *et al.*, 'Science and Technology Indicators for the UK: Innovations in Britain since 1945', mimeo, 1981.

Totsuka, H., 'A Case Study on the Transition from Piecework to Measured Daywork in BLMC', *Annals of the Institute of Social Science, University of Tokyo*, No. 22, 1981, 5–42.

Towse, A., 'The BL Incentive Scheme in Perspective', mimeo, Nuffield College, Oxford, 1982.

TUC, *Automation and Technological Change*, London: Congress House, 1965.

TUC, *Employment and Technology*, London: Congress House, 1979.

Turk, J., 'Work and Unemployment', in D. Shepherd, J. Turk, and A. Silberston (eds.), *Microeconomic Efficiency and Macroeconomic Performance*, Oxford: Philip Allan, 1983.

Turner, H. A., Clack, G., and Roberts, G., *Labour Relations in the Motor Industry*, London: George Allen and Unwin, 1967.

Turner, J. A., 'Computers in Bank Clerical Functions: Implications for Productivity and Quality of Working Life', unpublished Ph.D. thesis, Columbia University, 1980.

Ulman, L., 'Collective Bargaining and Industrial Efficiency', in R. Caves (ed.), *Britain's Economic Prospects*, Washington DC: Brooking Institution, 1968.

Utterback, J. M., 'Innovation in Industry and the Diffusion of Technology', *Science*, No. 183, 1974, 620–6.

Utterback, J. H., 'The dynamics of Product and Process Innovation in Industry', in C. T. Hill and J. H. Utterback, *Technological Innovation for an Expanding Economy*, Oxford: Pergamon, 1979.

Utterback, J. and Abernathy, W. J., 'A Dynamic Model of Product and Process Innovation', *Omega*, Vol. 3, 1975, 639–56.

Voss, C., 'Production/Operations Management: A Key Discipline and Area for Research', *Omega*, Vol. 12, No. 3, 1984, 309–21.

Walton, R. E., and McKersie, R. B., *A Behavioral Theory of Labor Negotiations*, New York: McGraw-Hill, 1965.

Waterson, D., *Economic Theory of the Industry*, Cambridge University Press, 1984.

Weekes, B., Mellish, M., Dickens, L., and Lloyd, J., *Industrial Relations and the Limits of the Law*, Oxford: Blackwell, 1975.

Weinstein, P., 'The Featherbedding Problem', *American Economic Review, Papers and Proceedings*, Vol. 54, No. 2, 1964, 145–53.

Wenban-Smith, G. C., 'Factors Influencing Recent Productivity Growth: Report on a Survey of Companies', *National Institute Economic Review*, 101, Aug. 1982, 56–67.

Wheeler, H., and Weikle, R. 'Technological Change and Industrial Relations in the United States', *Bulletin of Comparative Labour Relations*, No. 12, 1983, 15–35.

Wilkinson, B., *The Shopfloor Politics of New Technology*, London: Heinemann, 1983.

Williams, K., Williams, J., and Thomas, D., *Why are the British Bad at Manufacturing*?, London: Routledge and Kegan Paul, 1983.

Williams, R., and Moseley, R., 'Technology Agreements', mimeo, Technology Policy Unit, University of Aston, 1982.

Williamson, O. E., *Markets and Hierarchies: Analysis and Antitrust Implications*. Glencoe: Free Press, 1975.

Williamson, O. E., 'The Organisation of Work: A Comparative Institutional Assessment', *Journal of Economic Behaviour and Organisation*, Vol. 1, 1980, 5–38.

Williamson, O. E., 'Efficient Labor Organisation', mimeo, SSRC Conference on Economics and Work Organization, University of York, Mar. 1982.

Willman, P., 'Leadership and Trade Union Principles: Some Problems of

Management Sponsorship and Independence', *Industrial Relations Journal*, Vol. 11, No. 4, 1980, 39–50.

Willman, P., *Fairness, Collective Bargaining and Incomes Policy*, Oxford University Press, 1982a.

Willman, P., 'Opportunism in Labour Contracting: An Application of the Organisational Failures Framework', *Journal of Economic Behaviour and Organisation*, Vol. 3, 1982b, 83–98.

Willman, P., 'The Organisational Failures Framework and Industrial Sociology', in A. Francis, J. Turk and P. Willman, *Power, Efficiency and Institutions: A Critical Appraisal of the Markets and Hierarchies Paradigm*, London: Heinemann, 1983a.

Willman, P., 'Bargaining for Change: A Comparison of the UK and USA', Durham University Occasional Paper, No. 2, 1983b.

Willman, P., 'The Reform of Collective Bargaining and Strike Activity at BL Cars, *Industrial Relations Journal*, Vol. 15, No. 2, 1984, 6–18.

Willman, P., 'Industrial Relations Consequences and Implications of the Introduction of Microelectronic Technology', report for the Department of Employment, London: HMSO, 1986.

Willman, P. and Cowan, R. 'The Impact of Autotellers on Bank Staff Numbers', in M. Warner (ed.), *Microprocessors, Manpower and Society*. Aldershot: Gower, 1984.

Willman, P. and Morris, T., 'Union Growth and Membership Diversification: The Case of BIFU', paper to British Universities Industrial Relations Association Conference, mimeo, London Business School, 1985.

Willman, P., and Winch, G., *Innovation and Management Control: Labour Relations at BL Cars*, Cambridge University Press, 1985.

Wilson, D. C., Butler, R. J., Cray, D., Hickson, D. J., and Mallory, G. R., 'The Limits of Trade Union Power in Organisation Decision-making', Vol. 20, No. 3, Nov. 1982, 322–42.

Wilson, D. F., *Dockers: The Impact of Industrial Change*, *British Journal of Industrial Relations*, London: Fontana, 1972.

Wilson, R. A., 'The Impact of Information Technology on the Engineering Industry', research report, University of Warwick Institute of Employment Research, 1984.

Winch, G. (ed.), *Information Technology in Manufacturing Processes*, London: Rossendale, 1983.

Winchester, D., 'Power and Authority in the TUC', in J. Purcell and R. Smith (eds.), *The Control of Work*, London: Macmillan, 1979.

Winchester, D., 'Industrial Relations Research in Britain', *British Journal of Industrial Relations*, Vol. XXI, No. I, 1983, 100–15.

Windmuller, J. P., 'The Authority of National Trade Union Confederations: A Comparative Analysis', in D. B. Lipsky (ed.), *Union Power and Public Policy*, Ithaca: Cornell University, 1975.

Winsbury, R., *New Technology and the Press: A Study of Experience in the US,* London: HMSO, 1975.

Woodward, J., *Industrial Organisation; Theory and Practice*, Oxford University Press, 1965.

Woollard, T. G., *Principles of Mass and Flow Production*, London: Iliffe, 1954.

Wragg, D., and Robertson, J., 'Post-war Trends in Employment, Productivity, Output, Labour Costs and Prices by Industry in the UK', DE Research Paper 3, London: HMSO, 1978.

Wright, D., 'The Management Implications of On-Line Real Time Computer Applications: A Case Study of the Trustee Savings Bank', M.Phil. thesis, University of Liverpool, 1978.

Young, S., and Hood, N., *Chrysler UK: A Corporation in Transition*, London: Praeger, 1977.

Zeitlin, J., 'The Emergence of Shop Steward Organisation and Job Control in the British Car Industry', *History Workshop Journal*, No. 10, 1980, 119–37.

Zimbalist, A., (ed.), *Case Studies in the Labor Process*, New York: Monthly Review Press, 1979.

Author Index

Subject Index